NEUROBIOLOGICAL BACKGROUND OF EXPLORATION GEOSCIENCES

NEUROBIOLOGICAL BACKGROUND OF EXPLORATION GEOSCIENCES

NEW METHODS FOR DATA ANALYSIS BASED ON COGNITIVE CRITERIA

PAOLO DELL'AVERSANA
Senior Geophysicist
Milan, Italy

ACADEMIC PRESS

An imprint of Elsevier

Academic Press is an imprint of Elsevier
125 London Wall, London EC2Y 5AS, United Kingdom
525 B Street, Suite 1800, San Diego, CA 92101-4495, United States
50 Hampshire Street, 5th Floor, Cambridge, MA 02139, United States
The Boulevard, Langford Lane, Kidlington, Oxford OX5 1GB, United Kingdom

Library of Congress Cataloging-in-Publication Data
A catalog record for this book is available from the Library of Congress

British Library Cataloguing-in-Publication Data
A catalogue record for this book is available from the British Library

ISBN: 978-0-12-810480-4

For information on all Academic Press publications visit our website at
https://www.elsevier.com/books-and-journals

  Working together
to grow libraries in
developing countries

ELSEVIER   Book Aid
            International

www.elsevier.com • www.bookaid.org

Cover images

Top left and top right: Comparison of two different representations of the same turbidite channel
system. a) Single attribute map; b) RGB blend of three frequency magnitude responses (after Paton,
G.S. and Henderson, J., 2015. Visualization, interpretation, and cognitive cybernetics. Interpretation,
Vol. 3, No. 3 (August 2015); p. SX41–SX48. http://dx.doi.org/10.1190/INT-2014-0283.).

Bottom left: Shutterstock.com

Bottom right: RGB blend display showing the frequency change of a CT scan of the human brain.
Image courtesy of GeoTeric (private archive of GeoTeric).

Publisher: Nikki Levy
Acquisition Editor: Natalie Farra
Editorial Project Manager: Kristi Anderson
Production Project Manager: Edward Taylor
Designer: Miles Hitchen

Typeset by TNQ Books and Journals

Contents

III

BRAIN-BASED-TECHNOLOGIES AND BRAIN EMPOWERMENT

7. Brain-Based Technologies

8. Applications to Education in Geosciences

Preface

The innate attitude of human beings to wander and to explore the environment is well described by the English writer and traveler Bruce Chatwin (May 13, 1940—January 18, 1989). One of the central themes of his books is human restlessness. The crucial Chatwin's question was "Why do men wander rather than sit still?" Chatwin proposed his personal explanation of this dilemma in the book *Anatomy of Restlessness"* (posthumously published in 1996), a collection of essays, articles, short stories, and travel tales.[1] Using a *clipped, lapidary prose that compresses worlds into pages*,[2] he wrote that moving across long distances is "...an inseparable impulse of the central nervous system." In the same book, Chatwin discussed his thesis about nomadism and travelers. He stated that the *need to explore* is a basic instinct since the childhood. It is directly linked with our survival. Indeed, tracking down the path of animals was the first attitude of primitive men, to capture a prey or to avoid a predator.

Chatwin investigated another fundamental connection between the ancient human cultures and their environment in his famous book *The Songlines* (1987). The writer combined fiction and real stories for describing the trip to Australia that he had taken with a clear purpose: discovering the link between Aboriginal oral culture, language, environment, and nomadic travel. His thesis is that language started as song and that Aboriginal songs represent *mental maps* of the territory. Chatwin formulated the hypothesis that the "spreading of Songlines" across the land represents the general attitude of some ancient populations to create *imaginary maps*. The final intent is to preserve their culture. For the Aboriginals, as well as for other populations such as the Africans living in the Savannah, human beings and their environment are indissolubly linked. By "singing the land," men can recall the paths and can save them in a shared memory. Every specific song becomes a sort of detailed virtual map and creates a clear *mental image* shared by an entire community.

The opus of Bruce Chatwin helps me to introduce the central theme of this book: the deep neurobiological roots of exploration geosciences. In fact, the writer treated, from his original point of view, two important characteristics of human brain: our exploratory instinct and our ability to create

---

[1]The collection was brought together by Jan Borm and Matthew Graves, after the death of the writer.

[2]John Updike (1991) used these words for describing the prose of Chatwin.

mental maps. These qualities are deeply embedded in our *ancient brain* of primates (and of mammals) where our primordial instincts take place.

Exploration can be defined as "the act of searching information or resources." This definition implies both an intellectual and a physical activity. Famous seamen and adventurers such as Cristoforo Colombo and Marco Polo, as well as *intellectual explorers* such as Aristotle, Leonardo da Vinci, Charles Darwin, and Einstein, created the fundamentals of human knowledge, thanks to their strong impulse to explore. This attitude is supported by the intrinsic ability of the mammalian brain to recognize complex patterns, to create mental images and virtual maps of the environment, and to process and integrate these images into coherent models. Mental maps and images represent *the bricks of cognition*, as explained by the neurobiologist Antonio Damasio. He writes, "…the distinctive feature of brains … is their uncanny ability to create maps … But when brains make maps, they are also creating images, the main currency of our minds. Ultimately, consciousness allows us to experience maps as images, to manipulate those images, and to apply reasoning to them" (Damasio, 2010). Humans have exceptional skills to create, manipulate, and communicate images. They can transform these images into symbols. Using these symbols and saving them in some type of memory, human beings can create a phenomenon unique in the biosphere: the culture.

It is not difficult to agree that all these attitudes and abilities represent the fundamentals of exploration geosciences (as well as of other interpretative sciences, such as medical disciplines). In geophysical prospecting, for instance, every serious exploration project starts with a phase of planning and feasibility study. After acquisition, data are processed to produce images of the subsoil such as sections and volumes of physical properties. All that information is saved in computer memory, processed, and saved again. Geophysicists combine all these types of information. They create integrated models for extracting significant patterns and anomalies emerging from the background. Eventually, these can be associated with important resources such as ore deposits, water, hydrocarbons, and so forth. What happens in the brain when geophysicists and other geoscientists do their work? What parts of their cerebral cortex do they use when they create and interpret a seismic image of the subsoil? How can their brain recognize a meaningful signal, extracting it from a confusing background? How do the various lobes of their hemispheres cooperate for integrating different types of information? What ancestral neural connections turn on when they take the decision to start a new exploration project? What drives their choices when they select a model among many geophysical possibilities? Finally, what are the neurobiological roots of human exploratory instinct?

This book is an attempt to explore all these questions, borrowing the most advanced theories and experimental results from neurosciences. I will take care to distinguish consolidated data and theories from my ideas. First, I will introduce the fascinating (and reasonable) hypothesis that modern neurobiology can illuminate the key aspects of exploration geosciences from a cognitive point of view. With that goal in mind, I will summarize the basics of brain anatomy and physiology. However, it is not my intention to write a new book about the human brain. This is extensively described in many other books written by eminent cognitive scientists. Furthermore, this book is not a technical discussion about the geological and geophysical exploration methods. These are discussed in specific papers and books, and I would like avoiding useless redundancy. Instead, I will attempt a comparative study of human cognition and Earth disciplines. This field of research is relatively new and widely unexplored, although an increasing number of geoscientists and cognitivists pay attention to the links between their disciplines.[3] This growing interest is justified by the potential practical implications. In fact, scientific and technological developments in geosciences can be inspired, driven, and optimized by neurosciences. Over the past decade, new hardware, software, imaging algorithms, and interpretation methodologies in geosciences have been developed taking into account human cognition. I will indicate this type of technology with the expression "brain-based technology" (BBT).[4] This subject represents a central theme of my daily work as researcher in geosciences. I verified through my personal experience that analyzing, imaging, and interpreting geophysical information are more effective when I apply tools and methods based on cognitive criteria. In this book, I intend to explore and, possibly, clarify these criteria. Finally, I hope to convince the readers that a better comprehension of the functioning of our brain can improve our work in geosciences.

There is an additional aspect that motivates the comparative study of cognitive and Earth disciplines: brain empowerment. Geoscientists deal with complexity. They acquire and process multiscale and multidisciplinary data and try to create integrated models and complex images of

---

[3]In my previous book (Dell'Aversana, 2013), I investigated the links between geosciences, cognitive sciences, and epistemology. In several chapters, I mentioned also the work of other authors about the same topic. In this new book, I will focus the discussion mainly on the neurobiological background of exploration geosciences.

[4]The same expression of "brain-based technologies" is often used in the scientific literature to indicate various types of brain–machine interface. These technologies are commonly based on direct communication between the brain and computers, for assisting, augmenting, or repairing human cognitive or sensorimotor functions. Instead, in this book, the same terminology of "brain-based technologies" indicates methods for data analysis, interpretation, and teaching based on (and driven by) cognitive criteria.

xii   PREFACE

the Earth interior. The brain of exploration geophysicists and geologists is continuously stimulated to solve geological puzzles. Over the years, I have developed the fascinating idea that all this work on complexity could be intentionally addressed to brain empowerment. This possibility is supported by the modern concept of *neuroplasticity*. This refers to the potential that our brain has to reorganize itself by modifying the neural pathways during our entire life and not only in the childhood. In fact, individual connections within the brain are constantly removed, reinforced, or created, largely dependent on how we use the brain itself. I assume that exploration geosciences and BBTs can produce the *collateral effect* of improving both neurogenesis (formation of new neurons) and synaptogenesis (formation of new neural connections). I will explain this intriguing hypothesis in the final part of the book, after introducing the concepts of multimodal perception and expanded integration of geophysical data.

This book is an attempt to bridge neurosciences and geosciences with the intent to trigger new ideas and to inspire new researches in both domains. My goal is to stimulate a multicultural audience, including neuroscientists, cognitivists, geologists, and geophysicists. Finally, I hope that, after reading this book, students, researchers, and professionals will think about their respective scientific fields from a renewed, holistic point of view, based on the deep relationships between Earth disciplines and neurosciences.

References

Chatwin, B., 1996. In: Borm, J., Graves, M. (Eds.), Anatomy of Restlessness, Selected Writings, 1969–1989 (Posthumously published).

Chatwin, B., 1987. The Songlines. Franklin Press.

Damasio, A., 2010. Self Comes to Mind: Constructing the Conscious Brain. Pantheon, New York.

Dell'Aversana, P., 2013. Cognition in Geosciences: The Feeding Loop between Geo-disciplines, Cognitive Sciences and Epistemology. EAGE Publications, Elsevier.

Updike, J., 1991. Odd Jobs: Essays and Criticism. Knopf.

PART I

TWO (APPARENTLY) DIFFERENT WORLDS

CHAPTER

1

The Exploratory Brain

1.1 A SMALL-SCALE EXPLORATION CASE HISTORY

When I was a young geophysicist, I started working in a service company specialized in geophysical methods applied to civil engineering, environmental, and archaeological studies. The main part of my work consisted in acquisition, processing, and interpretation of geophysical data in railway tunnels for estimating their stability conditions. However, from time to time, I was asked to give a geophysical support to small-scale archaeological projects. Although I ventured into the fascinating domain of the archaeological exploration just few times, I was lucky to work in important locations, such as Pompeii and other historical areas in southern Italy. I learned to use the ground penetrating radar, often called geo-radar or briefly GPR (Clark, 1996; Conyers, 2004; Gaffney and Gater, 2003; Himi et al., 2016; Jol, 2008), combined with other geophysical techniques. Thus, I started the hard life of *field geophysicist*, doing many direct experiences in geophysical surveys in different sites of Italy. I remember that one of the most exciting "adventures" was a small geophysical project addressed to explore the interior of an ancient church built at the beginning of the 17th century in the historical center of Bari (Italy). The problem that I was asked to solve was theoretically simple, but very complex in practice. Based on ambiguous and fragmentary information extracted from ancient documents, the monks knew about the presence of rooms and/or tunnels located below the floor of their church. Unfortunately, nobody was able to confirm either the reliability of the old documents or the underground distribution of the cavities. An additional problem, very difficult to solve, was detecting the exact location of the entry (the trapdoor) of the underground cavities. In fact, the floor of the church was very ancient and precious. For that reason, the monks asked me to detect the original entrance of the cavities and to allow the access from the top. They wanted to avoid any large excavation for limiting the damages to the floor of their church. There was an additional, good motivation for detecting the small entrances with extreme accuracy: reducing the risk of structural collapse. In fact, opening a new way of access different from the original one and entering from the roof of the underground rooms would have affected the stability of the building itself.

We decided to explore the subsoil of the church using the GPR. I would like to recall just some basic information about geo-radar, for who has not any familiarity with this type of geophysical method. GPR uses radar pulses to image the subsurface. It works in the high-frequency range (from megahertz to gigahertz). Thus, resolution is extremely high (from meters to centimeters), but the investigation range is necessarily limited. This is because of the rapid attenuation of high-frequency electromagnetic waves, especially in conductive materials. For these reasons, the use of GPR is mainly directed to civil engineering, archaeological, and environmental applications. A GPR transmitter emits electromagnetic energy into the medium to be investigated. This can be the ground, but it can also be an artificial structure, like the wall of a building. When the electromagnetic waves encounter a boundary between materials characterized by different electric properties, they may be reflected, refracted, or scattered back to a receiving antenna. The signal detected by the receiver can be displayed in real time on the computer screen while the acquisition proceeds. These data can be interpreted immediately, when fast detection of buried objects is required. Alternatively, these data can be analyzed and interpreted later in the office. In the latter case, a complex processing workflow can be applied to enhance the signal-to-noise ratio, to convert the response from time to depth domain, and so forth. Similar to reflection seismic (but using a different type of waves), stack and/or migrated GPR sections (or volumes) represent the final output.

Before starting our survey inside the church, we collected all the available data about the building, including ancient documents, approximate descriptions by the monks, and fragmentary indications by local people. Then we tried to compose the puzzle of this fragmentary "a priori information." This was useful for limiting the investigation area, for defining an accurate acquisition program, and for optimizing the recording parameters. In this project, we planned to use two antennas having central frequency of 500 and 300 MHz, respectively. That choice was due to the expected size and depth of our targets. In fact, based on the available initial information, we expected to find the top of the rooms/tunnels at depth of less than 1 m below the ground. As I said, the access to these tunnels was expected to have a very small size (the monks talked about possible small trapdoors of about half meter width). Using an antenna of 500 MHz represented an ideal compromise between resolution and depth of investigation. Moreover, using also a 300-MHz antenna allowed us extending the exploration depth down to 4 to 5 m. That was a good setup for detecting relatively large rooms and tunnels and, at the same time, for detecting their small entrance with sufficient accuracy. With my team, I defined a dense network of reference points covering the floor. Finally, we started our survey using the 500-MHz antenna first.

I admit that, moving the antenna on the floor of the ancient church and looking at the GPR display in real time represented a unique experience for me. I was conscious that the target was *something* with high historical value. What did the tunnels represent? Were there any secret underground paths or something else? What did they hide? Could we discover ancient objects, relics, artifacts, paintings, old coins?

I think that geophysical prospecting addressed to archaeological exploration is one of the most exciting fields of applied geosciences. In fact, it represents a perfect combination of technology, science, and historical culture. Moreover, ability to capture weak signals, curiosity, enthusiasm, impulse toward the discovery, and serendipity are all instinctual attitudes necessary for a successful exploratory process. I felt these sensations all together while I was moving the antenna on the church floor.

Similar emotions motivated me to work for 12 h a day when I performed a similar geophysical survey in the area of Pompeii, just few months later. Back then, I was not conscious about that mix of feelings. Honestly, I did not try to investigate the nature of my emotions. I was just happy to do that work: a perfect integration of technical background and instinctual attitude to exploration.

1.2 MACABRE DISCOVERY

After 1 h of GPR surveying, we detected several large electromagnetic *anomalies* showing their top a few decimeters below the floor of the church. Fig. 1.1 shows two of these strong anomalies exactly as they

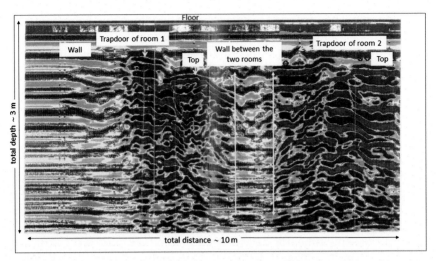

FIGURE 1.1   Ground penetrating radar anomalies detected from the floor of the church using the 500-MHz antenna (original display of 1990).

appeared on the original display of the GPR instrument back then (February 1990).

Although strong linear noise affects the quality of the image, two main anomalies emerge from the background. They are caused by the reflections and by many reverberations (black regions in the figure) of the electromagnetic waves at the top of two underground rooms separated by a thick wall. When the electromagnetic waves arrive at the top of the rooms, the high dielectric contrast between the background materials and the empty rooms creates strong reflections. The top of both the rooms clearly appears less than 1 m below the floor. Furthermore, the entrances (trapdoors) of the rooms are clear.

Looking at this real-time display it was easy for me to detect the optimal points for accessing into the rooms. Thus, we "sacrificed" just a small area of the precious floor of the church, performing an excavation in a very limited area. After a few minutes, I entered into the room 1 (see the figure). Of course, I was perfectly aware that I was not entering into the tomb of a pharaoh, but I was equally excited. I illuminated the dark room using a flashlight from the top. Unfortunately, no spectacular treasure appeared to me, but just a lot of ancient tiles, probably a residual of an old floor. I continued to descent prudently, followed by a monk, using the original stairway made with old stones.

Only when we were at the bottom we realized what that room was. On the wall in front of the stairway, there was a series of seats excavated in the rock. In the 17th century, people used to place the cadavers on these seats for draining their corporeal fluids, which were gradually collected by an underlying cavity. This operation was aimed at facilitating the decomposition of the dead body.

In summary, we discovered an ancient death chamber. Unfortunately, I did not explore the other rooms and/or tunnels detected during our GPR survey. The monks were satisfied to know, thanks to our work, the accurate location of every single trapdoor. Consequently, they decided to continue the direct inspection of the caves by themselves. I will never know the content of the other rooms. However, I am sure that I made the most macabre discovery of my carrier of exploration geophysicist.

1.3 LESSONS LEARNED

After that survey in the ancient church, I continued my job as an expert in GPR for another couple of years. Being a young geophysicist, I had all the necessary energy for traveling continuously, working in different environments, facing with extremely variable problems, and exploring various types of targets. In the following years, I was involved in different fields of geosciences, including research in seismology, volcanology, and, finally, hydrocarbon industry. However, I have never forgotten that story in the ancient church of Bari. I learned three very important lessons from it and from my other experiences.

First, despite the specificity of that geo-radar survey, it summarizes several general geophysical and geological aspects. These are planning, acquisition, processing, imaging, mapping, pattern recognition, and integration of information. These represent the fundamental "bricks" in the practice of exploration geosciences, independently from the physical nature and the spatial scale of the target. This can be a small death chamber under a church or a large magma chamber under a volcano, an archaeological artifact 1 m below the ground or a giant hydrocarbon reservoir at 4 km below the sea floor. These are different exploration objectives, but they share and imply a common methodological background.

Second, whatever the final target is, exploration geoscience is based on the application of crucial features of human cognition. These include our unique competence to plan the future and create expectations about it; the ability of our brain to process heterogeneous information, to create mental maps and virtual images of the reality; the human attitude to recognize significant signals, anomalies, and patterns of information and to integrate and save them in a shared memory. All these cognitive functions have been deeply studied in neurobiology. They take place in several interrelated areas of the neocortex, which is the most recently developed part of the brain.

Third, our rational cognition could not exist without the necessary support of basic emotions. Among these instinctual components of our mind, there is the impulse to explore the environment, imposed by the need to find the resources for our survival. Modern neuroimaging techniques, complemented by clinical and psychological approaches, allow us understanding, at least partially, the anatomy and the physiology of these ancient mental functions that make the human brain unique.

In this book, I intend to describe all these aspects, investigating the deep links existing among our activities of exploration geoscientists, our rational neocortex of *Homo sapiens*, and our ancestral brain of mammals.

References

Clark, A.J., 1996. Seeing Beneath the Soil. Prospecting Methods in Archaeology. B.T. Batsford Ltd., London, United Kingdom.

Conyers, L.B., 2004. Ground-penetrating Radar for Archaeology. AltaMira Press Ltd., Walnut Creek, CA, United States.

Gaffney, C., Gater, J., 2003. Revealing the Buried Past: Geophysics for Archaeologists. Tempus, Stroud, United Kingdom.

Himi, M., Pérez-Gracia, V., Casas, A., Caselles, O., Clapés, J., Rivero, L., August 2016. Non-destructive geophysical characterization of cultural heritage buildings: applications at Spanish cathedrals. First Break 34 (8), 93—101.

Jol, H.M. (Ed.), 2008. Ground Penetrating Radar Theory and Applications. Elsevier.

Cognitive Geosciences

2.1 INTEGRATING RATIONAL AND EMOTIONAL BRAINS

During the main part of the past 2 million years, the human species evolved in an environment resembling today's African savannah, or in other wild lands. Our ancestors lived in small nomadic groups of hunter-gatherers. Not before 10,000 years ago, a very recent time compared with our evolutionary history, just a limited part of the human population started becoming sedentary agriculturists. The brain adapted to the problems, which our hunter-gatherer ancestors encountered in their environment: finding a refuge, capturing a prey, escaping from a predator, discovering a spring, and finding the resources for surviving day by day. The size of our crania grew, to host increasingly large brains able to face the complexity of a wild environment. Nowadays, many of us live in huge and crowded towns, using (not in every case) the so-called *rational mind*. We apply technology and, hopefully, we adopt *rational* choices for optimizing the cost-to-benefit ratio in our daily activities. However, our ancestral and *irrational* brain works in the background using our primordial synaptic connections, influencing in some way the majority of our activities, including science.

Eminent scientists have described exhaustively the neurobiological basis of primordial emotions (Panksepp and Biven, 2012). Despite their work, many other scientists, but also common people, are still reluctant to admit that a rational activity such as science is largely driven by our ancestral brain. In the Earth disciplines, the practical sense of geologists, geophysicists, engineers, managers, and businesspersons pushes the efforts of everybody toward the development of new technology. Many of us think that these efforts must be rational, limiting the collateral effects of "dangerous" emotions. This approach could be defined as "radical rationalism." In this book, I will show that this point of view is unrealistic. Furthermore, it is misleading. I will show that women and men develop and apply technology using a complex cooperation of their primordial brain with the recently developed neocortex. Ignoring that cooperation and forgetting the ancestral origins of our mental attitudes could be a big mistake. Instead, investigating the complete neurobiological background of our scientific activity can be useful for driving the technology development itself and, finally, for improving our scientific and practical results.

What are, or should be, the crucial aspects of the "good" science and technology? I support the thesis that knowledge should be based on, and created by, a community of fully integrated brains. From a neurobiological point of view, that integration can be obtained throughout three complementary processes. The first two are related with each individual brain:

1. Improving the neural connections between the two cerebral hemispheres and between their constituent lobes (Fig. 2.1).
2. Improving the neural connections between the "rational" and "emotional" brains, respectively, the cerebral cortex and the subcortical nuclei (Fig. 2.2).

These two processes imply an extreme simplification of the concepts of "rational" and "irrational." However, they track the main road of an ambitious program of brain empowerment and development of "brain-based technology" (BBT). This type of technology is here intended as hardware, software, procedures, and workflows based on cognitive criteria and consistent with the key features of human brain.

Cerebral Hemispheres

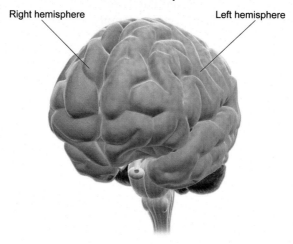

FIGURE 2.1   Cerebral hemispheres. *Image in the public domain, Blausen gallery, 2014.*

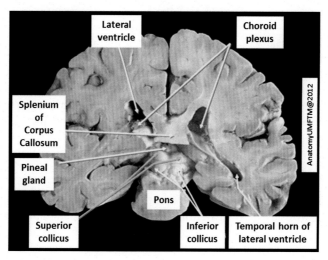

FIGURE 2.2   Transversal section showing the location of some parts of the "subcortical" brain, such as the superior and inferior collicus (or colliculus). *Image in the public domain, modified by Dell'Aversana, P., July 2015. The brain behind the scenes — neurobiological background of exploration geophysics, First Break 33, 41–47.*

The third process of the integration program, abovementioned, consists in improving the link and the cooperation between the multitude of brains forming an organized community[1] (the community of geophysicists, for instance).

The neurobiological background of these three processes of integration need to be explained in detail. This is my goal in the following part of this book, with the support of experimental results and theories borrowed from modern neurosciences. For that reason, I will provide a synthetic description of the main features and functions of the central nervous system (CNS). Of course, I invite the readers to deepen my descriptions (necessarily limited) using the wide scientific literature available on this matter. I will take care to provide the reader with continuous indications about the most recent references. Because of the strong link between geosciences and cognitive sciences, in the following I will refer to my approach (that integrates Earth disciplines and cognitive sciences) using the expression "cognitive geosciences." A specific sector of this novel discipline is "cognitive geophysics."

2.2 HUMAN AND MAMMALIAN BRAIN: AN OVERVIEW

The CNS consists of the brain and the spinal cord. The basic architecture of the brain evolves progressively during the entire life through the combination of genetic and environmental factors. Genes provide a sort of initial imprinting. However, even before the birth, environment influences whether and how these genes are expressed. Brain's architecture is similar for all the human beings in terms of physiology and distribution of the main anatomical parts. Of course, every individual brain is different from any other, because of the extreme variability of the specific experiences and the infinite possible combinations of genetic factors.

The study of the brain involves at least the following main fields: physiology (description of neural electrical activity, metabolism, neurotransmitters, receptors, etc.), anatomy (description of the main anatomical regions of the brain), and related functions. Of course a complete discussion about the brain includes many other fields, including molecular and cellular neurosciences, neurochemistry, study of neural circuits and large neural systems, cognitive and behavioral neurosciences, and

---

[1] In my previous book (Cognition in Geosciences, 2013) I introduced the concept of "semantic system": this is a combination of people, information, and (eventually) technology aimed at transforming sparse information into coherent concepts (models, theories, ideas, etc.).

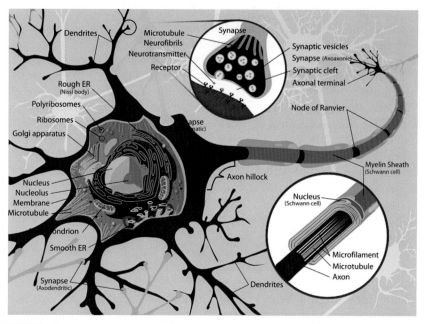

FIGURE 2.3 Diagram of neurons or nerve cells. *ER*, endoplasmic reticulum. *Courtesy: Mariana Ruiz Villarreal. Image released into the public domain.*

affective and clinical neurosciences, with the support of branches such as computational neurosciences, neuroimaging, social neurosciences, and so forth. Such a complete discussion is out of the scope of this book.[2]

2.2.1 Neurophysiology

Neurons (Fig. 2.3) represent one of the two main classes of cells composing the brain.[3] About 10^{11} neurons generate more than 10^{15} connections, which are responsible for the impressive complexity and, at the same time, for the uniqueness of every human mind. Neurons are very special cells. However, they have the key features of the other cells. The main anatomical element is the cell body that includes the cell nucleus; Golgi apparatus; and other organelles such as mitochondria, ribosomes,

[2] Among the others, a good reference for a scientific discussion about brain is Kandel et al. (2012). Additional interesting books accessible also to nonexperts in neurosciences have been written by Damasio (2010), Edelman (1987, 1992), Panksepp and Biven (2012). Images, atlas, neuroanatomy tutorials, and detailed descriptions of the brain can be found at the links included in the list of websites at the end of this chapter.

[3] The other type consists of glial cells.

microtubules, and so forth. The special feature of neurons is that they are involved in transmitting, receiving, and processing information. For performing these complex functions, neurons have a highly specialized structure. This allows them sending signals to other neurons in the form of electrochemical pulses called action potentials. The signal is transmitted by means of an axon. This protoplasmic fiber extends from the cell body and projects to other areas of the brain itself or to distant parts of the body. The electrochemical stimulation received from other neural cells arrives to the cell body, or soma, of the neuron, propagating through the dendrites (the input fibers). The transmission happens through the synapses. Neurons can be active (firing) or not (not firing), and synapses can be strong or weak. The strengthening of a synapse is a crucial factor for learning and memorizing.[4] In this case, the term *strength* must be intended as "ease of firing and thus ease of activation of the neurons downstream" (Damasio, 2010, p. 227). The signal-passing neuron is called presynaptic neuron and is localized on an axon; instead, the target neuron is called postsynaptic neuron. The chemical substances that transmit the signals across a synapse are called neurotransmitters. These play a fundamental role in regulating vital functions of the organism, including sleep, pain perception, body temperature, blood pressure, and hormonal activity. Basic emotions depend strongly on the amount of neurotransmitters produced and circulating in the different areas of the brain. This point is so important that it will be discussed with additional details in dedicated sections of this book.

Neurotransmitters can be broadly split into two main categories: small molecule neurotransmitters and the relatively larger neuropeptide neurotransmitters. The biogenic amines (dopamine, noradrenaline, serotonin, and histamine) belong to the first group. Instead, examples of neuropeptide neurotransmitters are corticotrophin, beta-endorphin, somatostatin, and vasopressin. Neurotransmitters are stored in synaptic vesicles that are clustered beneath the membrane in the axon terminal. This is located at the presynaptic side of the synapse. The life cycle of a neurotransmitter can be summarized in the following main phases: first, the neurotransmitter is synthesized in cell body, and then it is packaged into vesicles and successively released. Finally, it diffuses, metabolizes, and activates the postsynaptic receptors (Fig. 2.3).

Connected neurons form circuits and connected circuits form regions with an architecture that is topographically organized. The spatial distribution of neural circuits is ideal for *mapping*. These *maps* of interconnected neural groups are activated on two-dimensional (2D) surface sheaths stacked in layers forming the cerebral cortex. Other maps are

[4] This idea, nowadays fully accepted and experimentally confirmed, was intuitively introduced by Donald Hebb in the mid-20th century.

formed on other types of neural spatial organizations called nuclei (sort of grapes of neurons), such as geniculate nuclei and the collicular nuclei. These particular types of nuclei[5] are formed by 2D curvy layers and play a fundamental role in integration of multisensory information and in other basic vital functions (such as metabolism, visceral responses, and emotions). Some nuclei affect the functioning of endocrine and immune systems and are partially responsible of important aspects of consciousness. Nuclei allow the management of emotions and basic vital functions in those "primordial" brains with limited extension of cerebral cortex.

As I said, the cortex is formed not only by neurons (gray matter[6]) but also by other types of cells. The glial cells represent the main constituents of the CNS, with a total number that is about threefold that of neurons. They maintain homeostasis, form myelin, and assure support and protection for neurons. Glial cells do not have any direct control on neurotransmission; however, they help defining synaptic contacts and maintain the signaling abilities of neurons. There are various types of glial cells such as astrocytes, oligodendroglia, and microglia.

2.2.2 Anatomy and Related Functions

From an anatomic point of view, the brain is divided in two hemispheres (Fig. 2.1), linked by a complex system of fibers that form the so-called "corpus callosum" (see also Fig. 2.2). That anatomy corresponds (partially) to different capabilities commonly associated to each hemisphere. The left one senses the right part of the visual field, receiving the sensations and controlling the movements of the right part of the body. Moreover, it controls the main functions of the language. For instance, the Broca's and Wernicke's areas are specific zones of this lobe. They are related to speech production and the ability to understand the meaning of words,[7] respectively. This happens in the right-handed people. In that case, the left hemisphere represents the dominant one (the opposite happens in left-handed people). The right hemisphere controls the left

---

[5] These aggregates of neurons forming "nuclei" are not to be confused with the "cell nucleus" which are localized inside each neuron.

[6] Gray matter consists of closely packed neuron cell bodies forming specialized regions of the brain involved in functions such as muscle control, sensory perceptions, and so forth. Instead, white matter is neuronal tissue containing mainly long, myelinated axons (see Fig. 2.3 for neuron description).

[7] The function definition and the localization of Broca's and Wernicke's areas are more complex than this simplistic description. Recent studies indicate that functions earlier attributed to Wernicke's area occur more broadly in the temporal lobe and also in the Broca's area itself.

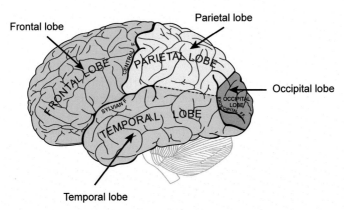

Frontal lobe

Parietal lobe

Occipital lobe

Temporal lobe

FIGURE 2.4   The cerebral lobes. *Image in the public domain. Modified from Gray, H., 1918. Anatomy of the Human Body. Lea & Febiger, 1918, Philadelphia; Bartleby.com, 2000. www.bartleby. com/107/, Revised by Warren H. Lewis.*

part of the body and of the visual field; it controls attention, spatial perception, and other perceptive/cognitive functions.

Both left and right hemispheres consist of several lobes (Fig. 2.4) that are related to different functions. Using an extreme simplification,[8] the frontal lobes are related to conscious thought and inhibition of some instinctual pulses. The parietal lobes are related to integration of sensory information from various senses, to manipulation of objects, and partially to visual-spatial processing. The occipital lobes are related to the sense of sight. The temporal lobes are related to the senses of smell and sound. However, we will see that the brain is extremely "plastic" and a rigid correspondence between specific areas and functions is not appropriate. The association cortices occupy a large portion of the cortex. These are located everywhere in the cerebral cortex where there are no early sensory cortices or motor cortices. They integrate all the sensory signals. For instance, the prefrontal cortices and the anterior temporal cortices are examples of (higher-order) association cortices. They combine information from several sensory association areas. The high-level cognitive functions such as language would not be possible without associative cortices.

The main anatomical difference between the brains of mammals and other vertebrates is the size. For instance, the brain of a mammal is 10 times as large as that of a reptile with the same body size. The distinctive feature of the human's brain is the massive expansion of the cerebral

---

[8] I will provide detailed functional/anatomical descriptions in different parts of the book. Here I am just introducing some of the key features of the brain for making clearer the following discussion.

cortex (the outer layer of neural tissue). Particularly large are the parts of the prefrontal cortex (involved in planning, working memory, motivation, and attention) and the parts involved in the vision. The cerebral cortex is characterized by high connectivity at variable scale: among single neurons, groups of neurons, and entire cortex areas. Moreover, the cortex is connected with the subcortical aggregates, or nuclei, mentioned earlier, through axon "projections." Important areas of the subcortical brain are the basal ganglia,[9] thalamus,[10] hypothalamus,[11] ventricles,[12] limbic system,[13] and the reticular activating system.[14] These are very similar in humans and in the other mammals and play a fundamental role in regulating basic emotions. This subcortical portion of the brain is often called "the ancestral or ancient or primordial brain" because it is developed before the cortex during the evolution of mammals. Of course, also human beings have this ancestral brain working in background behind the "rational" activity of the neocortical lobes (the newer portion of the cerebral cortex). Without it, we humans would have no emotions such as motivation, desire, expectations, rage, fear, care, etc. The subcortical brain of humans and that of other mammals look extremely similar. Instead, the two neocortices show huge differences. The human neocortex is much bigger and more folded-up than the neocortex of the other animals. The neurons in our ancestral brain have extended to innervate the larger human neocortex, where our high-level cognitive functions are generated, such as language, pattern recognition, and planning abilities. It means

[9] These are formed by the caudate nucleus, putamen, and globus pallidus, and they are involved in movement control.

[10] The thalamus has several fundamental functions, such as the relaying of sensory and motor signals to the cerebral cortex, the regulation of consciousness, sleep, and alertness.

[11] The main functions of the hypothalamus include control of appetite, sleep patterns, sexual drive, and response to anxiety.

[12] These are a number of cavities filled by cerebrospinal fluid, briefly CSF (e.g., glucose and electrolyte concentrations). This maintains and controls the extracellular environment and circulates endocrine hormones. CSF also surrounds the outer surfaces of the brain. It allows absorbing the shock caused by trauma.

[13] The limbic system consists of a series of nerve pathways incorporating other structures such as the hippocampus and the amygdala. The limbic system is involved in the control and expression of mood and emotion, in the processing and storage of recent memory; and furthermore, it controls the emotional responses to food.

[14] These nuclei receive input from most of the body's sensory systems and other parts of the brain (such as the cerebellum and cerebral hemispheres). Among the other functions, reticular formations have cardiovascular and respiratory control, wakefulness, overall degree of arousal, and consciousness.

that cortical and subcortical parts of the brain are extensively connected. For instance, the thalamus is a part of the ancestral brain that plays a fundamental role in the process of integration of information. It is located not too far from the center of the brain and is linked with many neural areas of the cortex. The thalamus–cortical connections are responsible for many of the superior cognitive functions of the brain, including consciousness (Edelman, 1987, 1992). The connections in the cortex are driven also by shape and spatial distribution of the neurons. As I said, these are organized in "maps" or aggregates. These maps have a fundamental function: they link the body receptors, like the photosensitive cells in the eye, with corresponding parts of the brain. Our extreme ability in integrating information, processing images, and developing complex concepts is intrinsically linked with the interactions between our cortical and subcortical neural aggregates. I will discuss in different parts of this book the links between these fundamental cognitive features with key aspects of geosciences.

Another two fundamental parts of the brain are the cerebellum and the brain stem. The first one is responsible for psychomotor function and coordinates sensory input from the inner ear and the muscles. Its main function is to provide accurate control of position and movement. At the base of the brain, there is the brain stem that links the cerebral cortex, white matter, and the spinal cord. It contributes to the control of breathing, sleep, and circulation. The CNS is connected to every point of the body by bundles of axons originating in neurons, commonly known as nerves. These transmit impulses from brain to body and vice versa and constitute the peripheral nervous system. A primordial part of it is the autonomic nervous system (working outside our volitional control) that plays a fundamental role in regulating emotions and feelings.

Of course, survival can be guaranteed only if the brain and the body communicate with high efficiency. This efficiency is obtained through chemical molecules such as hormones[15] traveling in the bloodstream. The brain has receptors for many types of hormones, such as the metabolic hormones insulin, insulin-like growth factor, ghrelin, and leptin. These can affect neuronal activity and certain aspects of neuronal structure significantly. Hypothalamus is one of the critical areas of the brain where some types of hormones are generated.

---

[15] Hormones are signaling molecules produced by glands in multicellular organisms. They are transported by the circulatory system to target distant organs for regulating physiology and behavior. Secretion of hormones may occur in many types of tissues. Endocrine glands represent the most important example. However, other specialized cells in various organs can secrete hormones in response to specific biochemical signals from a wide range of regulatory systems.

There is a fundamental connection between brain and immune system. One of the most important mechanism by which the immune system can influence the brain and behavior arises in the context of the inflammation process. Cytokines are small proteins that play a key role in cell signaling. They are produced by immune cells such as macrophages, B-lymphocytes, T-lymphocytes, and mast cells, and also by endothelial cells, fibroblasts, and various stromal cells. Cytokines can affect significantly the behavior of many types of cells. For instance, proinflammatory cytokines can access the CNS, interacting with the cytokine network in the brain itself. They can influence fundamental brain functions, such as neurotransmitter metabolism and synaptic plasticity. Significant effects can arise on neural paths and circuits regulating motor activity, motivation, anxiety, and so forth. Cognitive dysfunctions, depression, and other neuropsychiatric disorders can derive from the activation of specific types of cytokines (Capuron and Miller, 2011).

2.3 EXPLORATION GEOSCIENCES: AN OVERVIEW

After the brief overview about the brain, I introduce here some key aspects of exploration geosciences. Similar to the previous paragraph, this overview is aimed at creating the background for the following comparative discussion about geosciences and neurosciences.

Exploration geosciences include all those disciplines addressed to investigate the Earth interior, using variable methods and technologies in a wide range of spatial and temporal scales. The final objective is detecting and/or characterizing some type of *target*. For instance, there are geoexploratory methods aimed at discovering a hydrocarbon reservoir. Other applications are addressed to detect a magma chamber buried at a depth of several kilometers in the terrestrial crust. In a different context, geologists and geophysicists may work with archaeologists, for detecting artifacts buried at a depth of few meters or few centimeters. In every case, a robust geological background is required for interpreting all the complementary data. Sedimentology, paleontology, structural geology, petrology, mineralogy, and so forth concur with geophysics to explore the Earth interior. Moreover, other disciplines regarding economical evaluations, exploration risk assessment, geopolitical implications, logistic organization, and project management are commonly involved in the exploration workflow. This multidisciplinary approach is the norm, for instance, in hydrocarbon exploration. For that reason, exploration geologists and geophysicists should include in their background a basic knowledge of other disciplines complementary to geosciences.

Geophysical prospecting is a very important part of the exploration process. It is aimed at estimating physical properties of the Earth by applying a set of methods at sea or land surface, in boreholes, from air vehicles, and from satellites. Reflection and/or refraction seismic, gravity, magnetic, electric, and electromagnetic (EM) methods are commonly applied with different modalities depending on the nature of the target, the desired resolution, the available budget, and the geological and logistic conditions. There is an extensive literature about exploration geophysics (Telford et al., 1990; Sheriff, 1991; Turcotte and Schubert, 2002).[16] All geophysical disciplines are based on the measurement, analysis, and interpretation of some type of physical response produced by the investigated system.[17] Depending on the problem to solve, geophysicists can be interested in different properties of the medium, such as elastic, magnetic, EM, and thermodynamic properties. In fact, the different geophysical methods are generally sensitive to different properties. Consequently, combining the different methodologies into integrated workflows is the best approach for obtaining a comprehensive characterization of the system under study (Dell'Aversana, 2014). Geoscientists are used to distinguish seismic from nonseismic methods. This separation has just practical classification purposes.[18]

Seismic methods consist of a set of geophysical techniques based on the propagation of many types of seismic waves. They are commonly used to map the location of discontinuities of elastic properties, such as geological interfaces. A further basic distinction is between seismology and seismic prospecting. The first one studies mainly the earthquakes and the propagation of (naturally generated) elastic waves through the Earth; the second one includes a set of methods based on the analysis of elastic waves generated in the Earth using artificial means, such as dynamite charges, vibrating systems, or air guns. In both cases, the seismic waves produced by the energy source (either natural or artificial) pass through the Earth and are reflected, refracted, and diffracted at discontinuities of elastic properties. Part of the energy returns to surface and is received by instruments (geophones and hydrophones) placed on the ground, in boreholes, on the seafloor, and grouped in streamers of receivers dragged by dedicated vessels at sea surface. The predominance of seismic methods

---

[16] Geophysicists who already know the key steps of a geophysical workflow can skip this part. Students and nonexperts in this field can find here a useful summary about the main geophysical methods.

[17] This can be a natural geological system as well as an artificial system.

[18] In some cases, like for instance in hydrocarbon exploration, this sharp separation is partially justified by the predominance of seismic applications on other types of geophysical methods (electromagnetic, gravity, magnetic, etc.). Personally, I do not like this type of distinction because I think that all methods can provide important and complementary information.

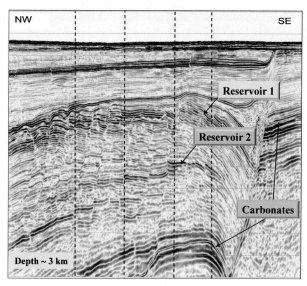

FIGURE 2.5   Example of seismic section. Total depth is about 3 km, length is about 10 km. *After Dell'Aversana, P., Colombo, S., Ciurlo, B., Leutscher, J., Seldal, J., 2012. CSEM data interpretation constrained by seismic and gravity data: an application in a complex geological setting. First Break 30 (11), 35–44.*

over other geophysical methods is generally because of their high accuracy and resolution. However, several intrinsic limitations affect also seismic data. For instance, the seismic response shows limited sensitivity to fluid variations in the rocks. Moreover, seismic data can show low signal-to-noise ratio in complex geological settings, such as in thrust belts and basalt-covered areas.

The final product of a seismic survey is generally an image of the Earth interior. This is obtained after applying a complex processing workflow (Yilmaz, 2001). This image can be a seismic section (in case of 2D surveys) or a seismic volume (in case of 3D surveys). Fig. 2.5 displays an example of section obtained through processing and imaging of seismic reflection data. Total length is about 10 km and total depth is about 3 km. The colors are related to reflection amplitude of seismic waves in correspondence of geological discontinuities (the vertical dashed bars in the figure are just markers). In this specific example, the main discontinuities correspond to the top of geological layers, such as two hydrocarbon reservoirs and a carbonate platform. The "reflection events" are commonly interpreted and tracked by exploration geophysicists picking them with different colors (see yellow, green, and cyan lines). The sharp lateral discontinuities of these reflections correspond to faults.

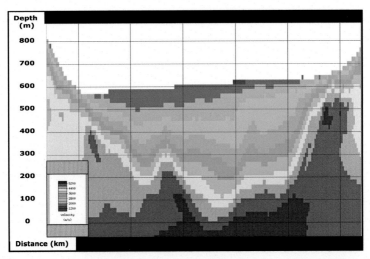

FIGURE 2.6   Example of seismic tomography section. Total horizontal distance is 7 km.

Fig. 2.5 is a typical example of seismic imaging used in hydrocarbon exploration. Geophysicists and geologists process and interpret seismic data with the final objective of creating a visual model of the Earth interior. Reflectivity of seismic waves represents a fundamental physical parameter because it provides relatively detailed information about the depth, shape, and size of the geological bodies of interest (such as a reservoir formation). An additional useful geophysical parameter is the velocity of propagation of the seismic waves into the different geological formations (Al-Chalabi, 2014). In fact, seismic velocity depends on the elastic properties that are generally different for the various types of rocks. For instance, the velocity of the compressional waves (V_p)[19] in carbonates is generally higher than that in shale formations. Based on that principle, another useful type of visual model, frequently used in geophysics, is provided by seismic tomography. This methodology allows mapping seismic velocity, retrieving it from the propagation times of the seismic waves measured by arrays of geophones. Fig. 2.6 is a 2D example of tomography image of seismic compressional velocity. In this case, colors represent the values (in m/s) of the propagation velocity. The red—yellow colors represent the high propagation velocity in a carbonate formation underlying the low-velocity sediments forming an alluvial valley (green—cyan—blue).

[19] Mechanical longitudinal waves, also called compressional waves, produce compression and rarefaction in the same direction of traveling through a medium. The other main type of wave is the transverse wave. In that case, the displacements of the medium are orthogonal with respect to the direction of propagation.

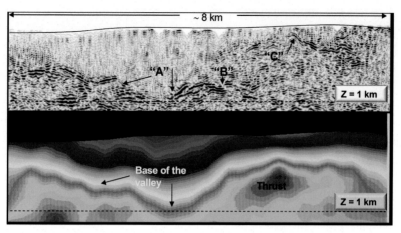

FIGURE 2.7   Combined imaging of a seismic section and the corresponding velocity model.

Of course, reflections and propagation velocities of seismic waves are reciprocally linked. A strong velocity contrast at the interface between two different geological formations will create an equally strong contrast in "seismic impedance" (product of bulk density and seismic velocity). The effect will be a reflection with high amplitude. Fig. 2.7 shows how seismic reflections are related to seismic velocities and roughly follow similar trends. The seismic section shown in the top panel is compared with the correspondent velocity model obtained through seismic tomography. Seismic quality is relatively poor because the data have been recorded in a complex geological setting (a thrust belt region of southern Italy). The signal is discontinuous and irregular. This is because of the complex geological trends causing equally complex ray paths, refractions, and diffractions. Some relevant reflections (A, B, and C) are indicated by the arrows. They are caused by the strong contrast in terms of velocity propagation in correspondence of significant geological transitions (like that from shale and carbonates displayed in blue–cyan and green–red, respectively).

Fig. 2.8 shows another example derived from seismic data analysis. It is a map, which shows the top of an important geological unit (a carbonate platform) obtained through interpretation of a 3D seismic data set. This type of display is frequently used by geoscientists; it shows essential structural information (depth variations, geological trends, faults, etc.) derived from analysis of geophysical data (interpretation, picking) and reflections detected in the 3D seismic cube. Maps like this allow building geological models of the Earth interior with high accuracy. These allow planning further geophysical surveys and/or deciding about the exact location of one or more wells. A good geological model must be based on

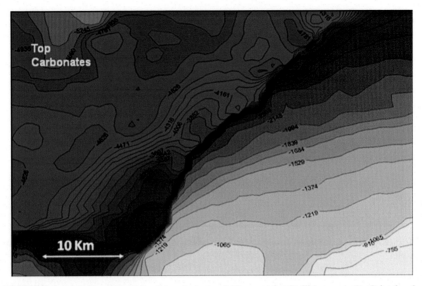

FIGURE 2.8   Map of the top of a carbonate platform. *After Dell'Aversana, P., Colombo, S., Ciurlo, B., Leutscher, J., Seldal, J., 2012. CSEM data interpretation constrained by seismic and gravity data: an application in a complex geological setting. First Break 30 (11), 35–44.*

high-quality seismic data that allow obtaining a satisfactory imaging of the subsoil. Imaging and mapping are fundamental steps of the exploration process.

Besides seismic methods, many other types of geophysical approaches are used in exploration geosciences. EM methods have a long tradition in geophysical prospecting due to their high sensitivity to some key physical properties characterizing the different types of rocks and fluids (such as electrical conductivity, dielectric constant, electrical chargeability, and so forth). An extensive discussion of the theoretical and applicative aspects of EM methods is given by Nabighian (1988, 2008). Another excellent overview is published by Zhdanov (2010). As a first general subdivision, we can distinguish between EM methods based on natural or controlled (artificial) sources. Among the first category, magnetotelluric (MT) methods are probably the most frequently used methods in deep geophysical prospecting of the Earth's crust and mantle. It estimates the Earth's EM impedance by measuring naturally occurring EM waves over a very broad frequency range. In hydrocarbon exploration, geophysicists use frequencies between 10^{-4} and 10^4 Hz with the objective to define the main geological trends down to a depth of several kilometers in the crust.

MT data are sensitive to electrical resistivity contrasts, which allow defining many geological features of interest in exploration. For instance, MT has been very useful in thrust belt exploration in some regions of southern Italy. Here a highly resistive carbonate oil reservoir is sealed by more conductive formations (prevalently clay and shale rocks). The main limitation of MT method is linked with its intrinsic low resolution, especially when it is applied in the low frequency range. This limitation affects all the EM methods operating at low frequencies.[20] It can be partially overcome by using "artificial sources," such as long electrical dipoles, controlled source power, and waveform. Among these methods, the most frequently used method in hydrocarbon exploration (but not the only) is the marine controlled source EM technique. In this approach, "a horizontal electric dipole excites both vertical and horizontal current flow in the seabed, maximizing resolution for a variety of structures" (Constable and Srnka, 2007). The Earth's response to EM fields is measured by electrical and magnetic receivers, commonly deployed on the seafloor. The data can be used to determine a multidimensional model of the subsurface resistivity. Such a model may be geologically interpreted to indicate the possible presence of hydrocarbon saturated layers.

Many other nonseismic methods are used in exploration geophysics. These measure different physical properties of the rocks, including mainly (but not exclusively) electrical, magnetic, and density properties. The most commonly applied methods are geoelectric, long-offset transient EM method (Strack, 1999), airborne EM, induced polarization (Nabighian, 1988, 2008), gravity (Fairhead, 2015), full tensor gravity, magnetic (Li and Krahenbuhl, 2015) methods, and so forth.

All these methods show their strength especially when they are properly combined into integrated multiparametric models. This integration can be obtained through complex workflows and algorithms that I will explain in the following chapters and in the appendices. Now, my intention is just to show that many different physical parameters characterizing the Earth interior can be corendered in the same display. The result is a multiphysics model that supports geological interpretation much better than any individual model based on one single parameter. Fig. 2.9 shows an example of corendered imaging of seismic, gravity, and EM information (Colombo et al., 2014). Panel A shows two seismic cross sections extracted from the 3D seismic cube. Panel B shows the corresponding velocity models. These highlight the different geological units, including the salt layer (the first red layer from the top). Panel C shows the density model obtained from gravity data. This is a smoothed spatial distribution of density; it provides useful information about the main trend of the basement rocks (the deep red layer). Finally, panel D shows

---

[20] Among the electromagnetic methods, ground-penetrating radar (or simply georadar, or GPR) mentioned in Chapter 1 has very high resolution because it works in the MHz–GHz range.

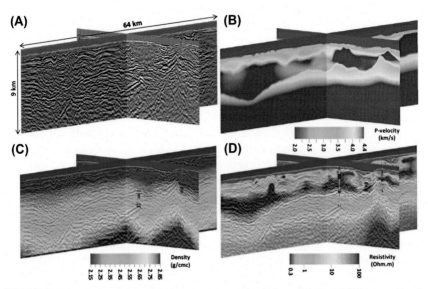

FIGURE 2.9   Example of multidisciplinary geophysical model. Panel (A): two seismic cross sections extracted from the 3D seismic cube. Panel (B): the correspondent velocity models. Panel (C): the density models obtained from gravity data. Panel (D): the resistivity models obtained from electromagnetic data. *Courtesy: Colombo, D., McNeice, G., Raterman, N., Turkoglu, E., Sandoval-Curiel, E., 2014. Massive Integration of 3D EM, Gravity and Seismic Data for Deepwater Subsalt Imaging in the Red Sea. Exp. Abstracts, SEG 2014.*

the resistivity models obtained from EM data. They add significant information about the geometry of the salt formation, including interesting shallow details not properly revealed by seismic data. The reason is that the EM response is very sensitive to the strong resistive contrast between several geological formations. For instance, the salt and the conductive background show a resistivity difference of about two orders of magnitude. Fig. 2.9 shows that each geophysical data contributes to solve a sort of geological puzzle. The final geological model will derive from appropriate combination of all these pieces of information, driven by geological criteria and constrained by borehole data.

2.4 EXPLORATION FROM A COGNITIVE POINT OF VIEW

Stimulating studies and interesting results have been published about the impact of cognitive sciences in geosciences (Dell'Aversana, 2011; Dell'Aversana, 2013; Froner et al. 2013; Gleicher et al. 2011; Gras, 2008; Henderson, 2012; Kastens and Manduca, 2012; Paton and Henderson, 2015; Shipley, 2013). As anticipated at the beginning of this chapter, several authors (including myself) call this integrated approach "cognitive geosciences." An important motivation for studying geosciences also from the point of view of neurosciences and cognitive disciplines[21] is the following: many "high-level cognitive processes"[22] applied by geologists, geophysicists, and engineers in their daily work (consciously or unconsciously) are deeply studied and well known by neuroscientists in their own field, although with different purposes and objectives. In other words, geoscientists and cognitivists partially share the same *matter*, but from different points of view. These shared aspects include the innate exploratory impulse, the intrinsic human ability in recognizing complex patterns, creating mental maps and images of the real world, integrating heterogeneous information, and so forth. Despite the good level of understanding of these crucial aspects of human cognition, unfortunately, systematic studies about the neurobiological background of exploration geosciences are quite rare in the scientific literature. Thus, investigating the links between geosciences and neurosciences can represent the first step for a novel, stimulating field of study. To highlight these links, Fig. 2.10 shows a comparison between a simplified block diagram of a hypothetical geophysical explorative project (left panel) and a simplified scheme of some high-level cognitive processes (right panel).

In the left panel, the boxes indicate the key steps of the geophysical workflow, from survey planning to the presentation of results. The black and red arrows show the transitions from a step to the next. We see that the workflow is not necessarily a linear sequence of activities. In fact, also circular loops between different processes can arise at variable scale. The presence of circular loops in the workflow is caused by possible feedback between different input, output, and processes. For instance, a velocity model can be the output of a complex data analysis; however, the model itself can modify the original interpretation of the seismic data and, consequently, can trigger a new processing workflow. The result can be an updated velocity model, and so on. Consequently, it is difficult to find any sharp separation between interrelated process, such as imaging,

---

[21] In this book, I often use the terms "neurobiology," "neurosciences," and "cognitive sciences" in an interchangeable way. That use can be justified by practical purposes, but it is not completely correct. The term "neurobiology" refers specifically to the biology of the nervous system, whereas "neurosciences" include the entire science of the nervous system. Neurosciences are an interdisciplinary domain in which medicine and biology are strictly related to "cognitive sciences," such as computer science, engineering, linguistics, mathematics, medicine, psychology, genetics, epistemology, and physics.

[22] "High-level cognition" or "macro-cognition" indicates the cognitive processes studied at a level higher than neurons or neural populations, such as planning, making analogies, recognizing patterns, integrating heterogeneous information, inductive reasoning, and creative thinking.

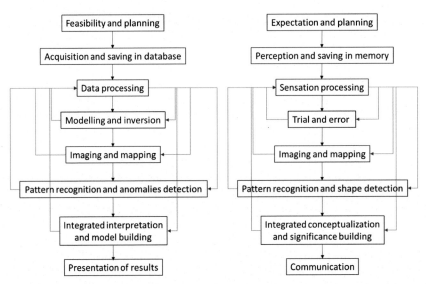

FIGURE 2.10   Simplified scheme of a geophysical workflow (left panel) and key cognitive steps (right panel).

inversion, and signal processing. For all these reasons, the figure is just an extreme simplification of a real geophysical workflow.

The right panel (sequence of key cognitive steps) is conceptually similar to the schematic workflow of the explorative project shown in the left panel. Effectively, a geophysical project reproduces at large scale what happens every day in our brain at a smaller scale. In fact, exploration represents an intrinsic attitude of humans (and of mammals in general). Consequently, it is reasonable to expect that the scientific community has developed geophysical approaches, year after year, consistently with the general structure of human cognition. This parallelism allows us explaining and describing the key exploration steps on neurobiological basis.

In the following chapters, I will investigate in detail the similitudes schematically showed in Fig. 2.10. I will start discussing the neurobiological background of the ancestral exploratory instinct in humans (and, more in general, in mammals). Then, I will focus the attention on several fundamental aspects involved in the exploration process itself that are strictly linked with each other. They are the attitude of our brain in creating "mental maps and images" of the world; our intrinsic ability in pattern recognition; and our capacity of integrating multisensory information for deriving complex concepts and interpretations of reality.

References

Al-Chalabi, M., 2014. Principles of Seismic Velocities and Time-to-Depth Conversion. EAGE Publications.

Capuron, L., Miller, A.H., May 2011. Immune system to brain signalling: neuropsychopharmacological implications. Pharmacol. Ther. 130 (2), 226−238. http://dx.doi.org/10.1016/j.pharmthera.2011.01.014. Epub 2011 Feb 17.

Colombo, D., McNeice, G., Raterman, N., Turkoglu, E., Sandoval-Curiel, E., 2014. Massive Integration of 3D EM, Gravity and Seismic Data for Deepwater Subsalt Imaging in the Red Sea. Exp. Abstracts, SEG 2014.

Constable, S., Srnka, L., 2007. An introduction to marine controlled-source electromagnetic methods for hydrocarbon exploration. Geophysics 72 (2), WA3−WA12.

Damasio, A., 2010. Self Comes to Mind: Constructing the Conscious Brain. Pantheon, New York.

Dell'Aversana, P., 2011. Creativity and innovation in complex informative systems. Applications in geosciences. First Break 29, 59−67.

Dell'Aversana, P., Colombo, S., Ciurlo, B., Leutscher, J., Seldal, J., 2012. CSEM data interpretation constrained by seismic and gravity data: an application in a complex geological setting. First Break 30 (11), 35−44.

Dell'Aversana, P., 2013. Cognition in Geosciences: The Feeding Loop Between Geodisciplines, Cognitive Sciences and Epistemology. EAGE Publications, Elsevier.

Dell'Aversana, P., 2014. Integrated Geophysical Models: Combining Rock Physics with Seismic, Electromagnetic and Gravity Data. EAGE Publications.

Dell'Aversana, P., July 2015. The brain behind the scenes − neurobiological background of exploration geophysics. First Break 33, pp.41−47.

Edelman, G.M., 1987. Neural Darwinism: The Theory of Neuronal Group Selection. Basic Books, New York, ISBN 0-19-286089-5.

Edelman, G.M., 1992. Bright Air, Brilliant Fire: On the Matter of the Mind. Basic Books, ISBN 0-465-00764-3. Reprint edition 1993.

Fairhead, D.J., 2015. Advances in Gravity and Magnetic Processing and Interpretation. EAGE Publications.

Froner, B., Purves, S.J., Lowell, J., Henderson, J., 2013. Perception of visual information: the role of color in seismic interpretation. First Break 31, 29−34. http://dx.doi.org/10.3997/1365-2397.2013010.

Gleicher, M., Albers, D., Walker, R., Jusufi, I., Hansen, C.D., Roberts, J.C., 2011. Visual comparison for information visualization. Inf. Vis. 10, 289−309. http://dx.doi.org/10.1177/1473871611416549.

Gras, R., 2008. The human factor in interpretation and visualization. In: 70th EAGE Conference and Exhibition, Extended Abstract, Workshop 6: 'Visualization: Is There Anything Left to Do?'.

Gray, H., 1918. Anatomy of the Human Body. Lea & Febiger, 1918, Philadelphia; Bartleby.com, 2000. www.bartleby.com/107/.

Henderson, J., 2012. Geological expression: data driven-interpreter guided approach to seismic interpretation. First Break 30, 73−78.

Kandel, E.R., Schwartz, J.H., Jessell, T.M., 2012. Principles of Neural Science, fifth ed. McGraw-Hill, New York, ISBN 0-8385-7701-6.

Kastens, K.A., Manduca, C.A., 2012. Earth and Mind II: A Synthesis of Research on Thinking and Learning in the Geosciences. The Geological Society of America. Special Paper 486.

Li, Y., Krahenbuhl, R., 2015. Gravity and Magnetic Methods in Mineral and Oil & Gas Exploration and Production. EAGE Publications.

Nabighian, M.N., 1988. Electromagnetic Methods in Applied Geophysics: Theory. Society of Exploration Geophysicists.

Nabighian, M.N., 2008. Electromagnetic Methods in Applied Geophysics: Applications, Part A and Part B. Society of Exploration Geophysicists.

Panksepp, J., Biven, L., 2012. The archaeology of mind: neuroevolutionary origins of human emotions. Nort. Ser. Interpers. Neurobiol.

Paton, G.S., Henderson, J., August 2015. Visualization, interpretation, and cognitive cybernetics. Interpretation 3 (3), SX41−SX48. http://dx.doi.org/10.1190/INT-2014-0283.

Sheriff, R.E., 1991. Geophysics. In: Sheriff, R.E. (Ed.), Encyclopedic Dictionary of Exploration Geophysics, third ed. Society of Exploration, ISBN 9781560800187.

Shipley, T.F., 2013. Structural geology practice and learning, from the perspective of cognitive science. J. Struct. Geol. 54, 72−84.

Strack, K.M., 1999. Exploration with Deep Transient Electromagnetics. Elsevier.

Telford, W.M., Geldart, L.P., Sheriff, R.E., 1990. Applied Geophysics. Cambridge University Press, ISBN 9780521339384.

Turcotte, D.L., Schubert, G., 2002. Geodynamics, second ed. Cambridge University Press, ISBN 0521666244.

Yilmaz, Ö., 2001. Seismic Data Analysis: Processing, Inversion, and Interpretation of Seismic Data. SEG Books.

Zhdanov, M.S., 2010. Electromagnetic geophysics: notes from the past and the road ahead. Geophysics 75 (5), 75A49−75A66.

Web References

https://images.nimh.nih.gov/public_il/.

http://library.pnca.edu/images/medicalimages.

http://library.med.utah.edu/WebPath/HISTHTML/NEURANAT/NEURANCA.html.

https://en.wikipedia.org/wiki/File:Complete_neuron_cell_diagram_en.svg.

WikiJ. Med. 1 (2), 2014. http://dx.doi.org/10.15347/wjm/2014.010. https://upload.wikimedia.org/wikiversity/en/7/72/Blausen_gallery_2014.pdf.

BRIDGING THE GAP

CHAPTER

3

Exploration

Neurobiological Background of Exploration Geosciences
http://dx.doi.org/10.1016/B978-0-12-810480-4.00003-9

3.1 GEOPHYSICAL PROBLEMS

The key word in exploration geosciences is "seeking." I have already remarked that exploration geophysicists and geologists look for many different types of *target* (water, oil, minerals, archaeological artifacts, etc.). However, despite the extreme variability of objectives and geological scenarios, the different types of exploration targets share one key aspect. This is the presence of *special patterns* and in particular, the evidence of some type of *anomaly*. The same concepts of *special patterns* and *anomaly* are fundamental also in many fields of medicine, including medical diagnosis and brain imaging.

What is an anomaly? The definition is not as immediate as we could think. However, it is sufficient to enumerate several examples of anomalies for clustering the distinctive features of an anomalous signal. In geophysics, we have gravity, magnetic, electric, electromagnetic, and seismic anomalies. In biology, we can detect a congenital vertebral anomaly or coronary artery anomaly or a genetic anomaly or a functional anomaly in cellular metabolisms. Moreover, we can also observe anomalies in physics, such as an unexpected effect in the interaction between subatomic particles or a weak gravitational effect caused by the interaction between two black holes happened 1 billion years ago.

All these anomalies have in common the fact that they produce some "irregular pattern" (an image, a sound, etc.) emerging from a "regular background." Consequently, the question of "anomaly detection" is strictly linked with the questions of "imaging and pattern recognition." Moreover, it is clear that an anomaly has a physical (or biological) meaning if we have a good comprehension of the background. In geosciences, this concept means that if we want to have some chance to detect significant anomalies we need to know the geological frame. For instance, a geophysical signal can be considered "special" in some way because it indicates the presence of a hydrocarbon reservoir. However, we can perceive that signal as an anomaly if and only if we have a reliable geological model to be used for interpreting our data. As a rule, a robust geological model is fundamental for performing an interesting discovery, unless we are extremely lucky. Whatever the target is, the searcher behaves like a sort of "hunter of anomalies" who seeks for special patterns. Sometimes these special patterns emerge clearly from the background; sometimes they are hidden and confused in a noisy context; sometimes they are just weak signals that can be reinforced through a special processing workflow (frequency filtering, stacking, deconvolution, and integration of multiple responses).

The practice of exploration geophysics includes examples of all these types of anomalous patterns. Great part of the geophysical work consists of detecting, processing, enhancing, imaging and, finally, interpreting geophysical patterns and anomalies. Fig. 1.1 (Chapter 1) shows an example of strong geo-radar anomalies. The irregular, high-amplitude reverberations (in black, in the figure) clearly emerge from a background characterized by linear signals with lower amplitude (prevalently in red and yellow). The same figure suggests that a geophysical anomaly can include particular signals related to specific targets. For instance, the trapdoors of both rooms produce particular reflection patterns, with a specific shape well localized in the space. The top of the first room (on the left) shows a curved shape resembling the vault. In this example, the difference between anomaly and background is clear, probably even for a nonexpert in interpretation of geo-radar data.

This example represents a successful exploration case at small scale, where the target was detected in real time (during the acquisition itself). No particular processing workflow was necessary for improving the signal-to-noise ratio. The anomalies emerged from the background with sufficient "strength" to allow us detecting the precise location of the targets. Unfortunately, this lucky scenario is not very common in exploration geophysics. For instance, in hydrocarbon exploration the target is often localized at depth of several kilometers in the Earth's crust, below thick and complex sedimentary sequences. Sometimes the geological setting is complicated by the presence of basalts and salt bodies. These geological bodies increase the complexity of the wave field propagation and degrade the final seismic image. In such difficult circumstances, exploration geoscientists must be able to interpret ambiguous signals. The patterns of interest are confused in a noisy scenario. Fig. 2.7 (Chapter 2) is an example of poor seismic data where the seismic reflections (in the top panel) are irregular and discontinuous. In such conditions, the geological trend can be interpreted with the help of tomographic images (see the bottom panel of the same figure) but with substantial uncertainties. In the following, I will discuss how geoscientists apply sophisticated techniques of imaging enhancement (including tomography). These techniques are not very different from the approaches used in diagnostic medicine and in brain imaging.

In this chapter, I discuss the neurobiological background of human attitude to search for some type of target. This can be a hydrocarbon reservoir for modern geoscientists or a prey hunted by ancient hominids. This is not a simple metaphor. As claimed by eminent neuroscientists (Panksepp and Biven, 2012; Damasio, 2010), the seeking instinct originates in our primordial brain of mammals. Ancient "neural circuits" govern this basic instinct to explore the environment, to follow a weak track, to identify interesting patterns, and to interpret meaningful anomalies. In the following paragraphs, I will discuss this interesting matter. First, I will talk about some epistemological and psychological aspects of the processes of seeking and discovering. Then I will discuss the neurobiological aspects of these fundamental processes in relation to exploration geosciences. Differently from the small-scale exploration case history of Chapter 1, I will discuss two exploration examples that involved a large community of geoscientists for long time, in a wide geographical region. After summarizing the crucial steps of these exploration stories, I will provide my personal interpretation of the facts from an unusual cognitive and neurobiological point of view.

3.1.1 Seeking Anomalies: Epistemological and Psychological Aspects

The well-consolidated *scientific method* is based on the virtual loop between induction and deduction. Scientists infer models and theories about the world based on their experimental observations. On the other side, they program the future experimental work starting from assumptions and expectations derived from their actual theories. This is a self-feeding loop between theory and practice or, in a more general sense, between deduction and induction.

A third way of thinking not identifiable with either the inductive inference or the deductive process is *abduction*. This type of reasoning, introduced by Charles Sanders Peirce[1] (1839—1914), can be summarized as "the process of formation of explicative hypothesis." It is useful when the standard hypotheses fail in explaining unexpected facts or when we do not have any guess at all. This process starts when new events happen that are not consistent with our expectations. Sometimes, we call these special events with the name *anomaly*. As I said, geophysics can be considered the science of anomalies: a hydrocarbon reservoir, a magma chamber, and an archaeological artifact are targets that generate anomalous geophysical responses with respect to the background. Consequently, geophysicists and geologists are familiar with abduction, consciously or not. It represents a basic approach for triggering the seeking process and for making discoveries. The general structure of abduction can be schematized as follows.

[1] Peirce first introduced the term as "guessing."

1. We observe several events and, among these, we see an event "A" that is significantly different from our expectations (anomaly).
2. We do not find any "standard" hypothesis for explaining "A" using our current knowledge. Thus, we try to explain "A" assuming a new hypothesis, "H," that is different from any hypothesis ever done until now.
3. If "H" were true, then "A" would be immediately explained. Consequently, we have robust motivations for *suspecting* that "H" is true.

Following this scheme, we see that abduction is a powerful process for generating new ideas in the form of new hypothesis. For instance, sometimes the abduction-based reasoning triggers innovative concepts in the exploration workflow. For instance, in exploration geosciences, revolutionary ideas about the *petroleum system*[2] in a certain region can emerge from anomalous evidences (geophysical data or drilling results) not explicable with the current models. This can introduce new exploration themes and, finally, can lead toward new discoveries.

On the other side, a new idea is not necessarily a good idea. Consequently, abduction can fail more than a robust theory based on a long sequence of experimental results. The abductive approach is more risky than the inductive approach. If we assume an innovative hypothesis based on few anomalous events (abduction), we cannot pretend any absolute confirmation of it in future. Instead, if we observe many consistent facts, we can be confident that Nature is revealing some uniform structure like a physical law, for instance (induction). Despite its intrinsic risk, the ultimate *reward* of abduction can be very high. The story of hydrocarbon exploration includes several game-changing discoveries rewriting the scenario of hydrocarbon industry in entire geographical regions (Esestime et al., 2016). Sometimes these successes start from some type of abductive reasoning and not exclusively from inductive thinking based on effective evidences. Only subsequent data and further drilling results confirm the robustness of the initial hypotheses.

I remark again that abduction does not start from ordinary facts but from anomalous events. Standard science uses induction and deduction for consolidating a background of knowledge, whereas revolutionary ideas are based on abduction focused on anomalies (Kuhn, 1962). This means that to perform abduction we need to seek and, possibly, to discover something special emerging from the ordinary scene. Furthermore, we need to understand properly what constitutes the background to focus our attention on anomalies. Innovations and discoveries emerge from the flatness of the

---

[2] In exploration geosciences, the petroleum system consists of a mature source rock, hydrocarbon migration pathway, reservoir rock, trap, and seal.

common sense, when the "explorer" is able to follow a weak track in a noisy environment. That principle is true not only in the everyday life but also in exploration geosciences and, in particular, in hydrocarbon exploration. In fact, as anticipated in the previous paragraph, in many Earth disciplines there is an additional complication: geologists and geophysicists often deal with ambiguous and weak signals. Thus, the work of seeking is more difficult than in other fields of study. The following statement seems to be written exactly for geoscientists: "...The hunter could have been the first to tell a story because only hunters knew how to read a coherent sequence of events from the silent (though not imperceptible) signs left by their prey" (Ginzburg and Davin, 1980).

The ability to seek and interpret "speechless tracks" and to transform them into coherent Earth models represents an essential attitude of exploration geoscientists. They must discover significant signals, such as small amplitude anomalies, seeking in the mass of a million seismic traces. In other cases, geophysicists must recognize weak electromagnetic signals. In complex geological settings, where data quality is poor, seismic interpreters have the ability to pick coherent horizons, draw faults, and see thrusts. When they are successful, geophysicists are able to detect significant information where others see only noise.

Indeed, geoscientists work like hunters: they track their prey (hydrocarbon traps, water or mineral accumulations) looking for weak signals and clustering sparse information in a coherent view. Their objective is to find a *conceptual structure* (a hypothesis, a model, a theory, etc.) that can explain many small anomalies from a geological point of view. Sometimes, the work of many geoscientists can be compared to the activity of a detective. In fact, both categories of "seekers" use conceptually equivalent methods, a sort of self-feeding loop of abduction combined with inductive and deductive thinking. Finally, they have, in principle, the same final objective: the discovery.

This approach is commonly supported by advanced technology. Moreover, it is often based on *serendipity*. Serendipity is strictly linked with the abductive process and corresponds with the particular ability to find something while searching for something else. It represents a sort of pleasant surprise. It is widely recognized that serendipity can play an important role in the scientific search, although it is often ignored in the scientific literature. In fact, standard scientific approach and scientific thinking is based mainly on logic and systematic search. Instead, serendipity is based on the attitude to capture weak signals, small anomalies, and "speechless tracks." This is an instinctual ability rather than a robust scientific method. Many special "seekers" such as detectives, hunters, explorers, and, of course, geophysicists and geologists share that basic instinct. They are able to find order where normal people did not see anything but confusion and inconsistent information.

3.1.2 Gestalt: Perception of Patterns

Of course, there are important differences between a geophysicist and a detective or between a geologist and a hunter. A key difference, for instance, is that a geophysicist rarely finds "spontaneous" signals on the ground, like a hunter or a detective. Sometimes there are special geological evidences at surface suggesting the presence of a deep target. However, these "signals" must be interpreted and converted into information that has a geological meaning. This work of transformation is commonly difficult and requires scientific competence. In general, there is not any prey or killer leaving tracks in the scene of the geophysical domain. The anomalies that, eventually, lead to a discovery or to an innovative idea must be extracted via long preparatory work. This is generally complex and requires many steps, algorithms, and a robust theoretical background. In summary, before attempting any abduction, a geoscientist needs to predispose the abductive process itself. In Earth disciplines, the work is commonly based on geological knowledge. This represents the necessary background from which we extract anomalies and interesting patterns.

An additional difficult task is to distinguish between significant anomalies, outliers, and artifacts. In fact, data inevitably contain noise, and the processing itself can produce many "false" signals. This discrimination between what is important and what is misleading is one of the most difficult parts of the work of geoscientists. Modern technologies allow us accessing redundant data sets and to an unbelievable amount of information. The ability to filter the messages is required. Neither a specific technique nor an optimal approach exists for doing this. However, several approaches can be applied to improve the capacity to filter the real anomalies from the artifacts. Apparently, these qualities deal with art more than with science. In fact, the Earth response is often a matter of shape rather than a matter of absolute amplitude. A seismic interpreter does not analyze the amplitude of every individual trace at a given reflection travel time. He/she analyzes a sequence of traces that, all together, produce a pattern. As a hunter learns to recognize the tracks left by different animals, in a similar way a geophysicist can learn to recognize the shape of the signals.

There is a German word that expresses the concept of shape: Gestalt. M. Wetheimer, K. Koffka, and Köhler in the 1920s analyzed the importance of the global shape that determines the structure of our primary perceptive processes. They understood that perception on its own is not independent from the high-level cognition: we see what we expect to see; we capture the shape as global information depending on the context, based on our previous experience, our culture, our expectations, our viewpoint, and so on. These concepts were developed and grouped under a new psychological paradigm called "Gestalt theory." These theories are extremely close to many aspects of the geosciences: it is not difficult to verify that seismic interpretation, signal processing, electromagnetic inversion, gravity modeling, and stratigraphic correlations are questions of Gestalt.

In summary, the sensitivity to the shape and the attitude for pattern recognition built over years of experience, a sound technical background, the comprehension of the context, etc. are the fundamental requisites for distinguishing among useful and misleading signals. The process of "anomaly building" from the raw data, the decision about what is background and what is anomaly, together with the efforts made to discriminate between anomalies, outliers, and artifacts, is preparatory work that anticipates, not only in geosciences, abduction itself. For this reason, I call this process preabduction.

Preabduction can be a spontaneous process, but more realistically, especially in complex domains, the preparatory work for highlighting the anomalies has to be based on precise techniques. For instance, in geosciences, the appropriate representation of the available information represents the first step for a proper identification of the true anomalies and for distinguishing meaningful signals from noise and from artifacts. More generally, *representation* is a primary thinking process for cognition. The proper attitude to representation is not only a technical matter; but it also involves an esthetic sense for the structures and for pattern recognition. It is probable that an artistic spirit has additional chances to "feel" some interesting signal hidden in the background and to recognize the meaningful anomalies (Gestalts). This was the case with a champion of human intelligence, Leonardo da Vinci. He was simultaneously a genius in science, technology, and art. I suppose that Leonardo's way of thinking was based on a strong attitude to represent reality and ideas combined with abductive reasoning.

3.1.3 New Look: Patterns in the Frame of a Model

High-level cognition is much more than perception of some type of Gestalt. Consequently, exploration geoscience is much more than anomaly detection. In 1947, Bruner published his classic study "Value and Need as Organizing Factors in Perception." In this work, he demonstrated with a series of experiments that children perception is not just an immediate response of the senses but depends on interpretative factors. In other words, the perception of significant patterns depends (also) on the "models" that we already have in mind. For instance, Bruner showed that poor and rich children perceived the symbol of the dollar in a different way, larger in the first case. Bruner's observations represented the experimental basis of a new movement in psychology, the "New Look." Following that point of view, every object of our experience is not perceived as an absolute "Gestalt" but is partially constructed by our mind. That cognitive construction can be realized, thanks to our previous knowledge, expectations, emotional states, personal history, social status, and so on.

Coming back to geosciences, for instance, a geophysical pattern can be perceived in different ways depending on the current geological model. A person without any appropriate background does not perceive the anomaly in the same way as an expert exploration geoscientist. It is not only a question of different possible interpretations of the same data but also a question of different perception.

Gestalt and New Look theories are not in conflict. They take into account for complementary aspects of how we develop knowledge through the combination of bottom-up and top-down processes (data-driven and model-driven approaches, respectively). Our knowledge derives from the continuous mix of bottom-up recognition with top-down cognition. That continuous combination has strong impact in the exploration geosciences. The first implication is the following: exploration geosciences cannot be successful if they are based just on the compulsive seeking of anomalies. They must be driven by geological knowledge. Special patterns, such as seismic or electromagnetic anomalies, will be useful only after their interpretation in the frame of a robust geological model. This fundamental concept will be clarified in the exploration case histories discussed in dedicated paragraphs of this chapter.

3.2 NEUROBIOLOGICAL BACKGROUND

The "ancient" mammalian brain underlying the cortex extensively influences human "high-level" cognitive functions of the cerebral cortex (and our consequent rational behavior). This concept has been widely discussed and supported by the neuroscientist Jaak Panksepp. He writes "… certain intrinsic aspirations of all mammalian minds, those of mice as well as men, are driven by the same ancient neurochemistry" (Panksepp, 1998). Among the basic instincts, a crucial behavior for survival is the exploratory impulse. In all the mammals, including human beings, the activity of exploration and research has deep neurobiological roots. Panksepp states that a well-defined *neural system* drives that primordial behavior. That system includes the amygdala, nucleus accumbens, lateral hypothalamus, and other subcortical neural nuclei (Panksepp and Biven, 2012).

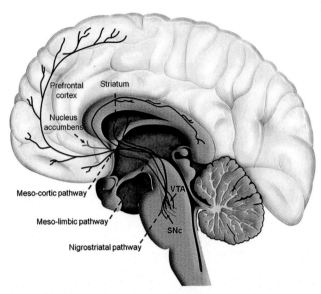

FIGURE 3.1   Dopamine pathway in the human brain. SNc, substantia nigra; VTA, ventral tegmental area. (Image in the public domain, under the terms of the Creative Commons Attribution License). *Credit: Arias-Carrión, O., Stamelou, M., Murillo-Rodríguez, E., Menéndez-González, M., Ernst Pöppel, E., 2010. Dopaminergic reward system: a short integrative review. Int. Arch. Med. 3, 24. http://dx.doi.org/10.1186/1755-7682-3-24. http://www.biomedcentral.com/1755-7682/3/24/. International Archives of Medicine.*

It energizes the frontal neocortical regions focused on all the primary needs and induces animals in exploring their environment (see Fig. 3.1 and explanations in the next paragraph). When this neural system is activated, mammals show particular interest for novelties. They try to capture opportunities from the environment (a prey, for instance) or are induced to escape from dangers (such as a predator). Panksepp's thesis is that this neural system has the ability to activate an instinctual emotional appetitive state. For that reason, this author labeled it as the "seeking system." All mammals share the same type of neural system in their typical seeking behaviors. These include goal-directed activities, learning, searching for resources, and looking for novelties (Alcaro et al., 2007). Panksepp says "… This system makes animals intensely interested in exploring their world and leads them to become excited when they are about to get what they desire. It eventually allows animals to find and eagerly anticipate the things they need for survival… When fully aroused, it helps filling the mind with interest and motivates organisms to move their bodies effortlessly in search of the things they need, crave, and desire. In humans, this may be one of the main brain systems that generate and sustain curiosity, even for intellectual pursuits. This system is obviously quite efficient at facilitating learning, especially mastering information about where material resources are situated and the best way to obtain them. It also helps assure that our bodies will work in smoothly patterned and effective ways in such quests." (Panksepp, 1998).

Experiments on rats showed the existence of specific "centers of pleasure" in the brain area of the seeking system. If the animals were given the way to self-stimulate these centers (for instance by pressing a lever that controls the onset of brain stimulation), they would continue until totally exhausted. Before collapsing, the stimulated rats moved about excitedly, sniffing and searching around, showing an obsessive-compulsive seeking behavior. Besides these pathologic behaviors, the seeking system of the brain is strongly involved in "positive" activities of search and exploration. This neural system is implicated when animals (including human beings, of course) search for significant connections between different types of information, facts, and events happening in the environment. The seeking system works in background when we try to create models about the world, meaningful hypothesis, novel concepts, and ideas connecting heterogeneous sensory data in a coherent way. The same neural pathways are active, although with different involvement of the neocortex, in the seeking system of a predator while searching for a prey and in the seeking system of a scientist while he/she is looking for a new theory.

Originally, this system was called "reward system" because it plays a key role in behaviors motivated by various types of reward. However, it is not the reward that makes animals feel euphoric but the search itself. In fact, the seeking activity triggers an expectation/appetitive state, a general positive feeling, and, finally, high levels of dopamine. All these effects drastically decrease after obtaining the goal. In other words, exploring seems to be even more important than finding. Of course, a successful seeking activity will produce pleasure and satisfaction. However, the brain—body cooperation is optimal before the success itself, during the research activity.

3.2.1 The Seeking System and the Dopaminergic Pathways

Dopamine is the molecule that plays the key role in the activation of this particular neural system. The chemical structure of this neurotransmitter consists of an amine obtained by removing a carboxyl group from a more complex molecule of L-DOPA. The term "dopaminergic (DA) pathways" is commonly used to indicate the neural pathways in the brain that transmit dopamine from one region of the brain to another (Fig. 3.1). Due to the role of dopamine, the seeking system is commonly associated to the "mesolimbic dopaminergic (ML-DA) system." In mammals, the majority of neurons containing dopamine (DA neurons) are clustered in three main mesencephalic[3] groups. These are located, without any sharp anatomical boundaries, in the retrorubral field[4] (cell group A8), in the substantia nigra[5] (cell group A9), and in the ventral tegmental area[6] (cell Group A10). Additional aggregates of DA neurons are located in the supramammillary region of the hypothalamus, areas of the dorsal raphe,[7] and periaqueductal gray.[8]

Being a neurotransmitter, dopamine influences synaptic activity. When it is transmitted through its pathways, it promotes the "release" of neural activity patterns. These trigger seeking behaviors such as loco-motion, sniffing, and head and body movements. Moreover, dopamine transmission supports the process of associative learning. Several neu-roscientists studied how this neurotransmitter can influence learning by enhancing synaptic plasticity (Centonze et al., 2001; Li et al., 2003; Huang et al., 2004) and by improving neural connections. In fact, dopamine can influence the activity of ion channels and/or can change the functionality of membrane receptors. It also regulates gene expres-sion, causing permanent synaptic changes (Greengard, 2001; Wolf et al., 2003; Nestler, 2004). Moreover, dopamine can promote generation of gamma rhythms.[9] This type of neural oscillation plays a key role in synaptic plasticity, memory, and other fundamental cognitive processes

[3] The mesencephalon, or *midbrain*, is a portion of the central nervous system associated with vision, hearing, motor control, sleep/wake, alertness, and temperature regulation. It consists of two main parts: the *tectum* (roof) and the *tegmentum*. The first one is the dorsal surface of the midbrain. It contains two pairs of bumps called *colliculi*. The posterior pair, called the *inferior colliculi*, has an auditory function; the anterior pair, called the *superior colliculi*, has a visual function. The tegmentum is the portion of the mesencephalon ventral to the tectum. It includes the *reticular formation* (or *tegmental nuclei*), plus *periaqueductal gray*, the *substantia nigra*, and the *red nucleus*.

[4] This is one component of the midbrain reticular nucleus.

[5] This is a brain structure that plays an important role in reward, addiction, and movement. The name derives from the dark color due to high levels of neuromelanin in DA neurons.

[6] This consists of a group of neurons rich in dopamine and serotonin neurons. It is located close to the substantia nigra and the red nucleus, near the midline on the floor of the midbrain.

[7] The dorsal raphe nucleus is located on the midline of the brain stem and is part of the raphe nucleus.

[8] The periaqueductal gray is the primary control center for descending pain modulation. It plays an important role in many vital functions linked with defensive, reproductive, maternal behavior, and with analgesia.

[9] Pattern of neural oscillation in humans with a frequency between 25 and 100 Hz.

(Paulsen and Sejnowski, 2000). Additional modulation effects produced by dopamine on neural activity are described by Alcaro et al. (2007).

3.2.2 The Seeking System and High-Level Cognition

When the seeking system is activated, the animal is in a state of high attention and euphoric engagement with the world. That state of psychological *arousal* is focused on both positive and negative aspects of the environment, such as discovering new resources or escaping from a possible danger. Instead, when this neural system is not activated or is damaged, exploratory and appetitive-approach behaviors are reduced or eliminated (Liu et al., 1998).

The proper functioning of the DA system is fundamental for promoting an active behavior of mammals in the majority of their activities related to instinctual exploration and to learning. Many neuroscientists generalized this statement for explaining several high-level human activities from a neurobiological point of view. Alcaro et al. (2007) write that human brain imaging data "... is beginning to highlight how important this system is in all varieties of appetitive human motivation, from the excitement of anticipating monetary rewards to the delights of love and music." Other neuroscientists (Blood and Zatorre, 2001) support this thesis. Indeed, Panksepp (1998) explains that the seeking system not only promotes a certain class of survival behaviors but also high-level activities not directly linked with instinctual abilities.

The normal (nor pathologically compulsive) activation of the DA pathways sustains curiosity even for intellectual goals. It facilitates learning, thinking in a creative way, discovering new connections between apparently separate events. This neurobiological process triggers a positive, self-feeding loop. When the seeking system works properly, we are motivated to interact with the world in a proactive/creative way. Consequently, dopamine circulation is maintained at the proper level, sustaining both physical and mental activities.

However, Panksepp remarks that the seeking system responds not only to positive incentives but also to all those situations where humans (and mammals, in general) must seek solutions in difficult scenarios. These can be represented by a predator attack or by a complex scientific puzzle. In other words, the DA pathways are involved also in situations of problem solving at extremely variable scale: in the daily routine as well as in the solution of complex scientific questions.

3.2.3 Connections With Other Neural Systems

Seeking system is just one among many other neural systems. Panksepp states that at least seven systems exist in the brain related to the basic emotions of all mammals, including human beings. These are the

SEEKING, ANGER, FEAR, PANIC-GRIEF, maternal CARE, PLEASURE/ LUST and PLAY systems (Panksepp, 1998).[10] Speaking about the last one, Panksepp writes, "A basic urge to play exists among the young of most mammalian species." This basic instinct is fundamental for allowing the animals increasing their physical fitness. However, it is extremely important also for consolidating social rules, developing new relationships, interacting with the environment, and thinking creatively. Using their system of play, young predators become expert hunters and young preys learn to avoid predators. This neural system is strictly linked to the brain centers deputed to somatosensory information processing within the midbrain. It is activated by certain reticular nuclei of thalamus and is especially connected with the sensorial system of the touch. Somatosensory data are projected up to the parietal cortex to process information coming from bodily sensations. Moreover, the same data are projected to thalamic nuclei that sustain a playful motivational state.

Clinical observations show that bilateral lesions to thalamic areas involved in the neural circuitry of the play system drastically reduce the impulse to play. The neocortex is not essential for triggering this basic instinct. Animals with their cortex partially removed still show a significant impulse to play. On the other side, the activation of the play system can influence significantly various high-level cognitive functions, such as planning and learning.

The main neurotransmitters involved in the activation of the play system are not only the opioids, but also the dopamine, acetylcholine, and serotonin. Although the desire to play decreases if the basic survival necessity is not satisfied, experimental evidences on laboratory rats suggest that the systems of play and the seeking system are often activated together. They cooperate for promoting many types of adaptive behavior, such as learning and social integration. These systems work also in the ancient mammalian brain underlying the human neocortex. Their cooperation is essential for triggering and supporting the ability to explore the environment, the attitude to work in team, and the capability to find creative solutions.

3.3 EXPLORATION, SEEKING, AND SPATIAL THINKING

Seeking activity implies moving across some type of space. This can be a *physical space*, if the target is a physical objective, such as water, food, and a refuge. Moreover, it can be a *conceptual space*, if the goal of the research is a creative idea or the solution to a theoretical problem. In any

[10] Panksepp uses capital letters for indicating these neural systems.

case, an additional distinctive cognitive aspect involved in active seeking behavior is *spatial thinking*. That aspect clearly emerges in exploration geosciences, as well explained by Kastens and Manduca (2012) and Kastens and Ishikawa (2006). Geologists, mineralogists, sedimentologists, petrologists, and geomorphologists must face with a huge range of natural and conceptual "objects," at variable spatial (and temporal) scale. That extreme variability influences significantly their cognition, especially in field geology. The authors highlight that expert geoscientists gain sophisticated projective and Euclidean spatial concepts through repeated practice. For instance, structural geologists develop advanced signal-recognition skills and a specific ability to capture "special shapes" (such as folded sedimentary layers, faults, and other geological discontinuities).

These attitudes increase over the time while moving across a complex landscape and trying to interpret it from a geological point of view. Mapping complex patterns, envisioning how a geological layer extends into the subsurface, interpreting the grain size distribution of sedimentary particles represent common challenges for the field geologist. He/she progressively develops the so-called "disembedding" attitude (Reynolds, 2012). This is a special ability "… which involves observing a complex scene (whether an outcrop, landscape, or map) … recognizing patterns, and isolating the important aspects (the signal) from distracting, nonessential ones (the noise)."

Moreover, geoscientists develop a special memory for recurring patterns, such as a particular sequence of sedimentary rocks, a special distribution of structural elements, a distinctive association of fossils and/or minerals. The activity of observing rock outcrops; measuring, classifying, discovering relationships among information; inferring the shape and the orientation of buried layers, over the time contributes to develop a special "exploratory brain" in the majority of geoscientists.

There is an additional special aspect that characterizes the activity of field geologists: they must be able to locate themselves on the geological or topographic map. In other words, different from other categories of "observers," geoscientists move inside the same space that they are studying. This implies a distinctive cognitive condition that requires a spatial thinking different from the ordinary.

Finally, Kastens and Ishikawa (2006) remark the importance of metaphorical use of spatial thinking. They suggest that geoscientists think in a different way from ordinary people because they often use "… a spatial dimension of a data display as a visual analogy to represent time." This mental approach "allows the geoscientists to reveal or highlight causal relationships. For example, Sclater et al. (1971) depiction of seafloor age versus depth in oceans of different spreading rates helped reveal the process of lithospheric cooling."

3.4 LINKS AND EXAMPLES

In the previous paragraphs, I tried to analyze the same crucial questions about the seeking activity from different points of views: epistemological, psychological, neurobiological, and, of course, geological and geophysical. It can be useful to schematize the key concepts of the discussion to search for possible inedited links.[11]

- A key aspect of exploration geosciences is searching for interesting signals and significant patterns in the frame of a robust geological model.
- The neuroscientist Jaak Panksepp teaches that the neurobiological roots of the human exploration instinct originate in the ancient neurochemistry of the DA system, combined with the activation of other neurotransmitters (glutamate, serotonin, and opioids).
- Recently, geoscientists and neuroscientists highlighted some peculiar aspects of spatial thinking in geologists, mineralogists, palaeontologists, and so forth.

Let us try to compose the jigsaw puzzle.

Panksepp's theory suggests, indirectly, that the neurobiological background of exploration geosciences could inhabit the seeking and the play systems of human brain. This is a reasonable hypothesis that could address further specific research programs. Many case histories, at different spatial and temporal scale, seem to support that hypothesis. The simple exploration anecdote discussed in the first chapter, represents just a small-scale example based on my direct experience. It shows how our rational mind and our emotions are perfectly combined when we search for some important target. From a neurobiological point of view, that combination could correspond to a neural link between neocortex and primordial subcortical nuclei forming our mammalian midbrain. As widely demonstrated by Panksepp, this neural connection is effectively realized in all mammals through the DA pathways. That neural link evolved over the time leading to the effective exploratory attitude of human beings. Over the past century, we combined our seeking instinct with advanced technology, creating scientific disciplines such as diagnostic medicine and exploration geosciences.

Nowadays, the increasing number of studies about geocognition and spatial thinking suggests that geoscientists use their seeking system in a very special way. In fact, they have a distinctive spatial (and temporal) perception by which they explore the environment differently from ordinary people. This special "seeking attitude" can create the conditions

---

[11] I remark that when we look for possible links between different conceptual domains, we activate our seeking system.

for making great discoveries. For instance, in hydrocarbon industry, many exploration successes can be properly explained only invoking both rational and emotional components, combined with the distinctive spatial cognition of geologists and geophysicists.

3.4.1 A Large-Scale Exploration Case History

Different from the small-scale exploration case history of Chapter 1, in this section I discuss an example of hydrocarbon discovery that involved, for many decades, a large community of geoscientists. It was a discovery that influenced the economic and social development of an entire geographical area. This example is instructive because it shows all the main aspects of the seeking activity addressed to a large-scale exploration target. These aspects include imaging and imagination, anomalous pattern recognition and integration of heterogeneous information, analogical thinking, and creative inferences. I will discuss in detail all these "exploration factors" in the next chapters. However, we can learn from now instructive lessons from this case history about the epistemological and neurobiological background of exploration geosciences.

3.4.1.1 Texas of Italy

Val d'Agri (Agri Valley) is a rift valley of Quaternary age filled with rubble—alluvial materials located in southern Italy, in the middle of the Apennine Mountain range (Basilicata). The Southern Apennines are part of the central Mediterranean orogenic belt formed by a sequence of thrust nappes[12] derived from the deformation of Mesozoic—Cenozoic domain. That deformation was caused by the Late Cretaceous—Quaternary convergence between the African and European plates. The geological units of the South Apennines form a thrust sequence from bottom to top that corresponds to an east-to-west transect in the original paleogeography. The mountains surrounding the Val d'Agri are (partially) made up of geological limestone formations of carbonate platform units (Apennine platform) overlapping the sedimentary sequences of an ancient sea basin (Lagonegro—Molise geological units). These consist of Triassic—Miocene carbonates, siliceous, marl, and siliciclastic rocks, related to the dismantling of the Apennine range and the consequent deposition of sediments in the foredeep.[13] Over the past century, the entire area and the subsoil have been intensively studied and explored for hydrocarbon research. Oil companies drilled many wells in the alluvial deposits of Agri Valley and

[12] In geology, a nappe or thrust sheet is a large sheetlike body of rock that has been translated for kilometers by tectonic forces from its original position.

[13] A foredeep is a geological basin filled with a thick accumulation of sediments derived from an orogenic belt during uplift.

on the surrounding mountains, crossing in depth all the main units of the Apennine sedimentary sequence.[14]

Val d'Agri is also called "Texas of Italy." In fact, in that region, the Europe's largest onshore oil field was discovered.

3.4.1.2 Beginning of the Story

The story of hydrocarbon exploration in Val d'Agri[15] started from the small and ancient village of Tramutola.[16] The interest for hydrocarbons derived initially from the curiosity of local people for the unusual physical phenomenon of hydrocarbon seepages visible near the village and already known since the Middle Ages. An oil (or gas) seep is a place where natural liquid (or gaseous) hydrocarbons escape to the surface along geological layers, or through fractures in the rocks. These phenomena generally occur above petroleum accumulation structures, indicating the possible presence of a deep hydrocarbon reservoir. The seepages of Tramutola are still visible near the small river now called "Torrente Fossatello," in east of the village. They were noted and described by Benedictine monks when they arrived in the Val d'Agri area in the Middle Ages. However, only after many centuries Tramutola seepages were scientifically studied and analyzed.

The systematic research of hydrocarbons in Val d'Agri effectively started at the end of the 19th century. At that time, the seepages of Tramutola were mentioned by Baldacci and Viola (1894), in the frame of a comprehensive geological study of the Southern Apennines. In 1902, Camillo Crema published the geological map of Tramutola with the title of "Zona Petrolifera di Tramutola,[17]" included in a brief technical report (Crema, 1902). This study was committed by the Inspectorate of the "R. Corpo delle Miniere" (mining department). It indicates clearly the increasing interest of the authorities in that region for a possible large-scale hydrocarbon discovery.

3.4.1.3 Oil Companies

In 1912, the Società Petroli d'Italia, signed the first contracts with local landowners for starting an exploration program in the area of Val d'Agri, with particular focus in the Tramutola zone. The initial exploration attempts did not produce any significant result. However, it contributed to

---

[14] Going from bottom to top, the sequence consists of the "Apulia" carbonate platform, the "Lagonegro–Molise" basins, the "Apennine" carbonate platform, and the internal oceanic to transitional "Liguride–sicilide" basin formations.

[15] An excellent summary of the geology of Val d'Agri and of the story of the discovery is provided by Van Dijk et al. (2013).

[16] This village is located on the southwest side of the "high" Agri Valley.

[17] Petroleum area of Tramutola.

increase the geological knowledge in the area and motivated further studies and researches.

The Italian oil company, AGIP (now Eni SpA), was effectively involved in the exploration of Val d'Agri since 1933, acquiring the research block of Tramutola. The location of the first exploration well was already defined in March 1933, and intensive geological studies continued in 1934. Moreover, in the 1930s, the new exploration methodology based on seismic prospecting was introduced in Italy and applied by AGIP. The young engineer Tiziano Rocco performed two seismic profiles in the Agri Valley. Unfortunately, the quality of the seismic data was very poor. In fact, the possible exploration target was not visible from the seismic sections. That pioneering geophysical survey showed an intrinsic difficulty: the geological complexity of that area of Southern Apennines represented an obstacle for imaging the subsoil by seismic prospecting. Indeed, the positioning of the first Tramutola well was based mainly on geological studies and surficial observations, without any significant support provided by seismic data.

3.4.1.4 The First Discovery

AGIP made the first discovery in 1937, drilling the well "Tramutola 1," located not too far from the seepages. It was a small and relatively shallow discovery: the main productive layers were found in arenaceous and calcareous intercalations at a depth between 190 and 270 m. However, it encouraged the exploration activity in the area. In the period between 1936 and 1943, AGIP drilled 47 wells. The Italian company found oil in 21 wells, gas in 3 wells, and oil and gas in 6 wells. However, the production was relatively small. Also in the other wells, the main productive layers were found at small depth, generally between 200 and 500 m.

Exploration in the area continued until 1959 when the last well (sterile) in Tramutola was drilled. Over the following years no significant exploration activity was performed. Tramutola oil field was effectively an exploration success, but geologists understood that it did not represent the main "hydrocarbon treasure" in that region.

3.4.1.5 Where Is the Real Treasure?

Since the first wells, the geological scenario emerging from the drilling results was extremely complex. Signorini (1939) analyzed the well data and recognized a "double" sequence of sediments formed in two completely different geological domains (platform carbonates and deep-sea basins). If those sediments were formed in different geological environments, how did they come in contact? The idea of long translational movements and tectonic overlap of large-scale geological units started to emerge. This concept was fundamental for changing the geological model of the Val d'Agri and for motivating the exploration interest toward deep targets. In 1949, C. Migliorini suggested the hypothesis of an extended anticlinal structure in the "Cretaceous and Eocene limestone formation underlying the chaotic outcrops[18]" of Tramutola (Migliorini, 1949). This new geological hypothesis focused the interest toward a new exploration objective, much deeper than the productive arenaceous and calcareous intercalations drilled by the Tramutola wells. The focus on that target was supported by further studies (Merla, 1952; Flores, 1955) and by analogies with the geology of other mountainous regions (in Northern Apennines and in Sicily).

In parallel with the work of Italian geologists, since 1952, the French geologists Glangeaud (1952), Grandjacquet (1961), and Caire et al. (1960) introduced new ideas about the formation of the Apennines. They described that mountain range as a complex thrust belt domain formed by extended eastward movements and superposition of entire geological units (carbonate platforms and basin units). On the other side, these innovative ideas could not be verified by the geophysical data available at that time. Seismic imaging had no sufficient resolution power, and it did not allow visualizing any deep geological structure. The new geological model was confirmed only many years later.

3.4.1.6 A Change of Paradigm

As it often happens, a significant step forward happened in correspondence of a critical conjuncture: with the oil crisis of 1973,[19] exploration in Basilicata restarted massively. In 1975, AGIP performed an extended exploration survey in Basilicata, using the new seismic technology. In 1979, AGIP and Montedison commissioned to "Ricerca e Interpretazioni Geofisiche" company a seismic survey in the area of Viggiano (in the Val d'Agri region). A few years later the company Petrex performed massive seismic acquisition in the Monte Alpi exploration block. The new seismic data allowed imaging the deep geological formations. However, the quality of the seismic images was still poor and the geological interpretation of the discontinuous seismic reflectors was difficult even for expert geophysicists. Despite their quality and reliability, all these new data allowed supporting the new ideas about the geological evolution of Southern Apennines. These were further developed by the "new generation" of Italian geologists (Scandone, 1967, 1968, 1969, 1971, 1972, 1975; Scandone and Sgrosso, 1964; Scandone and Bonardi, 1967).[20] A completely new picture of the Southern Apennines emerged after all

---

[18] My translation from the original Italian document.

[19] The 1973 oil crisis began in October 1973 when the members of the Organization of Arab Petroleum Exporting Countries proclaimed an oil embargo.

[20] In this brief overview, my reference list is necessarily limited. For a more exhaustive bibliography, see Van Dijk et al. (2013).

those studies and researches. Two different paleogeographic domains were identified: one was characterized by prevalent carbonate sedimentation, typical of shallow-water conditions, called "Apennine Carbonate Platform"; the other one was characterized by sedimentation typical of deep-water conditions, called "Lagonegro Basin," formed by a calcareous—siliceous—marls sedimentary sequence.

This regional reconstruction had a significant impact on the interpretation of Tramutola outcrops. Under the light of the new geological view, they were reinterpreted as parts to two distinct sedimentary series formed in two separate paleogeography areas, confirming the hypothesis of R. Signorini formulated since 1939. They came in contact only after long movements over geological times. The geological units involved in wide-range translations, such as entire carbonate platforms, are called "allochthones." The hypothesis of extended movements of geological allochthones forming the Apennines Mountains was progressively confirmed by the new deep wells drilled by AGIP in southern and central Italy.[21] Finally, Mostardini and Merlini (1986) provided a general synthesis of the new Apennine model, based on wells and geophysical data.

3.4.1.7 Modern Exploration

The period between 1970 and 1980 was characterized by intensive studies of surface geology combined with the interpretation of reflection seismic, gravity, and magnetic data. Combining all the available information, the majority of geologists and geophysicists agreed about the fact that all the outcrops are allochthones "floating" on a hypothetic, deep carbonate platform. The top of the deep carbonate platform (called "Apula") appeared as a discontinuous reflector in the best case. In many cases, it was invisible because of the attenuation and scattering of the seismic wave field by the complex overlying allochthones. The interpretation work of geologists and geophysicists was extremely challenging. Sometimes, it was driven by a significant amount of *imagination* rather than by real physical and geological constraints. The term "imagination" is here intended as a positive attitude of exploration geoscientists, which is their unique ability to imagine deep geological models from surficial evidences (outcrops) and weak signals (seismic reflections).

3.4.1.8 Finally the Big Discovery

Finally, all these scientific efforts culminated into the first exploration success. In 1981, the well "Costa Molina 1" found oil in the deep carbonate series, now called "Internal Apula Platform," confirming the validity of the new exploration theme. Exploration continued over the following

---

[21] That geological model was indicated in Italian as model of "completa alloctonia." It is fully in line with the plate tectonic theory.

decade. The wells "Mt. Alpi 1" (1988; operator Petrex, in joint venture with Enterprise Oil) and the well "Tempa Rossa 1" (1989; operator Fina, in joint venture with Lasmo and Enterprise Oil) confirmed the presence of extended oil reservoirs in the Val d'Agri area. The last decade of the millennium was characterized by further exploration through the drilling of appraisal wells ("Mt. Enoc," "Cerro Falcone," "Alli," "Tempa d'Emma," "Perticara," "Gorgoglione," "Tempa Rossa"). The main oil companies operating in the area performed extended seismic campaigns. 3D seismic surveys were operated by AGIP, in partnership with Enterprise Oil: the first one was the "3D seismic project of Monte Alpi" that covered a surface of 250 km^2; the second one was the "3D seismic project of Cerro Falcone" that covered a surface of 600 km^2. Moreover, innovative geophysical methods were applied in both Val d'Agri and Cerro Falcone areas, including long-offset seismic and electromagnetic methodologies (Dell'Aversana, 2001; 2003; 2014).

Based on all these geophysical surveys, integrated with the drilling results, the actual geological model includes two main structures of reservoirs. The first structure is below the Val d'Agri and is called "Trend 1" or "Val d'Agri Trend." It includes the sectors of Monte Alpi, Monte Enoc, Cerro Falcone, and Caldarosa. The second large-scale structure is called "Trend 2" or "Tempa Rossa Trend," drilled by Tempa d'Emma, Tempa Rossa, Perticara, and Gorgoglione wells. Production at the Val d'Agri field started in 1996. It reached about 83,000 oil barrels per day (b/d) in 2014. Production comes from the Monte Alpi, Monte Enoc, and Cerro Falcone fields. It is fed by 24 production wells and treated at the Viggiano oil center with an oil capacity larger than 100,000 b/d.

3.4.2 Lessons Learned

One century of exploration history in Val d'Agri shows all the crucial factors of a large-scale seeking activity that involved the local population, the international scientific community, an entire sector of hydrocarbon industry, managers, politicians, and administrators. I would like to summarize the lessons learned from the story, underlining the key words of the seeking process.

1. The hydrocarbon seepages at Tramutola represented the *anomaly* that triggered the interest of local people, geologists, researchers, and managers of oil companies toward a common *target*. However, at the beginning of the story, the nature of that target was not completely clear to everybody. Common people and institutions did not realize the potential implications of that natural phenomenon. It was just something anomalous, to be further investigated.

2. After that initial trigger, the innate attitude to explore the environment and to seek for natural resources was progressively enhanced by practical, scientific, and economic interests. *Expectations* increased in terms of scientific curiosity, business, and economic implications. A growing number of geological studies were committed in the area. Serious studies and the first wells improved the shallow geological model. This initial part of the story is a good example of how human knowledge can improve through the *self-feeding loop between seeking instinct and rational activity.*

3. The first oil discovery in Tramutola represented the preliminary *reward* of this initial exploratory phase. The seeking activity requires continuous resources; consequently, some type of reward is crucial for motivating and sustaining the efforts. Moreover, the expectation of a reward commonly induces an *appetitive state*. It is probable that a psychological condition of strong *expectation* addressed the geologists, local institutions, and private companies toward more and more ambitious exploration targets.

4. It is important to remark that, at this stage, the exploration activity was driven largely by the ability of geologists *to imagine* a model of the subsoil, just projecting the surficial information to depth and using the wells as local calibrations. *Imagination* and *spatial thinking* represented crucial attitudes.

5. An additional crucial factor of this exploration story is *analogical thinking*. In fact, new ideas about the model of Southern Apennines were derived also from *analogies and similitudes* with the geology of other mountainous regions of Italy and from the ideas introduced by French geologists. *Analogical thinking* is fundamental in the seeking activity and is complementary to *logical thinking*.

6. After the complete exploitation of the Tramutola field, exploration was targeted to a deep geological objective. Regional geological studies suggested that a significant oil production was possible only in deep geological formations. That hypothesis met initial skepticism. Drilling deep wells was too expensive and there were no actual evidences of deep reservoirs. However, the new geological ideas were confirmed by further studies. Moreover, new seismic technology allowed deeper *imaging* and new drilling technology allowed deeper wells. We can say that in this phase, the process of *imaging* supported the attitude of *imagination* and both sustained the seeking process. In other words, the exploratory instinct was supported by the ability of geologists to *imagine* the subsoil and by the ability of geophysicists to obtain *images* of the Earth interior. This is a synthetic description of the work of exploration geoscientists: it is a combination of instinctual attitudes (imagination) supported by rational approaches (such as imaging techniques).

7. An additional important remark is that both geological and geophysical information was progressively *reinterpreted* and *integrated* under the light of the new geological model of the Southern Apennines. This is a good example of how the process of *pattern recognition* is influenced by the context (geological model). In geosciences, this is the norm: the same data can be reinterpreted in a different way under the light of a new model. In other words, exploration geoscientists are able to combine, consciously or not, the basic principles of both *Gestalt* (perception of significant patterns) and *New Look Psychology* (interpretation of the same patterns in the frame of an expanded context).
8. New oil discoveries arrived in the 1980s. Finally, the wells Monte Alpi 1 (1988) and Tempa Rossa 1 (1989) confirmed the presence of huge oil field in the area of Val d'Agri. Additional discoveries were performed in the following years. All these exploration successes represent the final *reward* of the exploration activity performed by a large community of geoscientists.

3.4.3 Cognitive Interpretation

In my previous book (Dell'Aversana, 2013) I remarked how geosciences offer a special point of view for investigating high-level cognition. Sometimes, the practical activity of geologists and geophysicists illuminates crucial aspects of the functioning of our brain. Indeed, the large-scale exploration case history of Val d'Agri is an example of how the seeking system can work in a collective way.[22] When a multitude of brains are focused on the same target, as it happens in exploration geosciences, the innate attitude of human beings to explore the environment is exponentially enhanced. *Seeking* becomes something more than an individual instinctive pulse. It evolves into a social behavior that can produce scientific, cultural, and practical effects on entire communities.

I have been involved directly in the last part of the exploration story described in the previous paragraphs. In fact, I have been working for Enterprise Oil from 1996 to 2002 and for Eni from 2002 until now.[23] For many years, I have provided my small contribution in developing imaging methodologies and in the process of geophysical data interpretation in Val d'Agri. The main problem that I faced was to develop new

---

[22] Of course, this is just my personal interpretation from a cognitive point of view. I am confident that this enlarged view of that exploration case history can stimulate future researches in geocognition.

[23] I wrote this paragraph in March 2016.

geophysical approaches for imaging the deep Apula platform. This was almost invisible under kilometers of alluvial sediments and allochthones. I had the possibility to "feel" my personal *seeking system* in action while exploring the subsoil in Val d'Agri and in other regions of Southern Apennines. Today, I assume that the *DA pathways* of my brain were fully activated for finding new solutions to that difficult geophysical problem of target detection (Dell'Aversana, 2001, 2003, 2014).

I remember perfectly my feeling during the acquisition and processing steps of my geophysical work. It was not different from the emotions that I felt a couple of decades earlier, when I explored the subsoil under the floor of the old church in Bari (see Chapter 1), searching for old archaeological treasures. My *attention* was focused to capture weak seismic signals, indicating the possible presence of the top of the Apula platform. Moreover, *curiosity, enthusiasm, impulse towards the discovery,* and *serendipity* sustained my work in the field and during the work of data analysis. I was conscious that these are all instinctual attitudes necessary for a successful exploratory process, in the case of shallow geo-radar surveys as well as in deep hydrocarbon exploration.

During my geological studies and my geophysical work, I met personally some of the key geoscientists who developed the geological model of Val d'Agri and of Southern Apennines. In fact, I had the privilege to attend their lessons when I was a young student in geology and, many years later, to participate in their field trips, when I was a senior geophysicist. I like to think that I have worked with eminent geoscientists while their own seeking system was driving them in the field. I remember their ability to imagine the complex geometry of ancient carbonate platforms and basins buried under kilometers of sediments. I have seen their *spatial thinking* in action while they described the movements of allochthones using their hands or drawing simple sketches on the notebook. I remember their imagination when they extrapolated to depth the surficial observations. I cannot forget the ability of these geologists to link and correlate outcrops at distance of many kilometers, reconstructing the paleogeography of entire regions. They used their imagination in perfect symbiosis with their scientific background.

3.4.4 A Game-Changing Discovery: Thinking Out of the Box

I would like to mention another case history of exploration geosciences that shows how the seeking process can provide exceptional results when geoscientists and exploration managers think out of the box. In the summer of 2015, Eni discovered the Zohr supergiant gas field in the

Egyptian offshore.[24] This is the largest ever found in the Mediterranean Sea. Reading the official news and the technical papers published on specialized journals, everybody can understand the importance of this exploration success. It is based on the expertise and the technological innovation capacity of the entire company. However, additional "human" factors also contributed to the result. These include the ability to change a consolidated exploration paradigm in the Nile Delta. In fact, since the discovery of the Abu Madi Field in the Nile Delta of Egypt in 1967 performed by Eni (formerly AGIP), exploration in that area has been focused mainly on siliciclastic plays. In the 1990s, the exploration focus moved to deep-water Pliocene turbidite play fairways and, recently, to Early Miocene and Oligocene targets. In addition, carbonate plays have been targeted close to the cost but without any significant success. About four decades of exploration seemed to confirm the Eastern Mediterranean area as a clastic play basin. After drilling the well Zohr-1 in summer 2015, the exploration picture changed drastically. In fact, this well drilled a carbonate reef and lagoon buildup at the southern margin of an extended carbonate platform.[25] That result induced a significant revision of the previous ideas about the paleogeographic evolution of the Levantine Basin.[26] In fact, the presence of a carbonate buildup so distal from the actual Nile Delta requires the formulation of a new geodynamic model for that area. The Zohr carbonate buildup implies shallow-water conditions during the Early to Middle Miocene. Instead, prevalent deep-water conditions across the basin during the Neogene had been assumed[27] north of the Nile Delta.

A good summary of the geological frame of this area, the stratigraphy and the petroleum system, revised after the Zohr gas discovery and supported by seismic data, is provided by Esestime et al. (2016). These authors highlight that several crucial geological conditions converge for making Zohr a *perfect hydrocarbon discovery*. In fact, the buildup is a very

[24] The discussion of this discovery is necessarily shorter than the previous case of Val d'Agri for several good reasons. First, because it is more recent. Second, because obvious confidentiality reasons prevent to disclose technical and commercial details. Consequently, my description is confined just to the basic geological and geophysical aspects. Indeed, my intention is to use this case history for highlighting some crucial epistemological and methodological aspects of the process of exploration/discovery.

[25] The Egyptian subsidiary of Eni, IEOC, drilled a new play close to the Egypt/Cyprus border. They discovered a 628 m gross gas column, 430 m net pay and gas in place in Early Middle Miocene limestone formation, in a carbonate reef and lagoon buildup.

[26] Levantine Sea is the name of the easternmost part of the Mediterranean Sea.

[27] With the exception of the brief Messinian salinity crisis.

porous "reef facies" with a good internal connectivity; it is surrounded by late Miocene mudstones and capped by thick Messinian evaporitic rocks working as a good formation seal. The probable source for the biogenic gas in Zohr is represented by the Oligocene–Miocene mudstone surrounding the carbonate buildup.

Reformulating the regional geological model and addressing the exploration toward a deeper target, completely different from the traditional siliciclastic and turbidite plays, represented the crucial conceptual steps for thinking "out of the box." Nowadays, understanding all the geological factors is important for full exploitation of the Zohr discovery. Moreover, as it happened in other exploration successes (including Val d'Agri discovery) the new geological model can drive the exploration in the whole region seeking for other similar targets. In fact, the conditions of the Zohr gas play could be repeatable on regional scale in both Egypt and Cyprus. Indeed, "The opportunity exists to discover sufficient gas … to rewrite the play book for the development of the gas industry in the Eastern Mediterranean" (Esestime et al. 2016).

3.5 SUMMARY AND FINAL REMARKS

The successful exploration stories described (synthetically) in this chapter, combined with the most recent theories about the functioning of the "emotional brain" (affective neurosciences), help us to design the general structure of the geocognitive "seeking process." The block diagram of Fig. 3.2 summarizes the key features of this process. This scheme can be used for optimizing the cognitive potentialities of geoscientists during the exploration process and for driving the development of new brain-based technologies.

The two top boxes highlight the main neurobiological systems of the exploration activity. This is based on the strict cooperation between instinctual (left blocks) and rational (right blocks) aspects.[28] The neural seeking system described in this chapter works in symbiosis with the neocortical lobes of the brain to *trigger*, *sustain*, and *drive* the process toward some type of reward, such as a great hydrocarbon discovery.

3.5.1 Triggering the Process

Seeking is a basic instinct of all mammals. It can be more or less active depending on external circumstances (triggering causes) and/or on

---

[28] This is just an extreme simplification. In fact, as shown in Fig. 3.1, the Dopaminergic pathways go through the prefrontal cortex too.

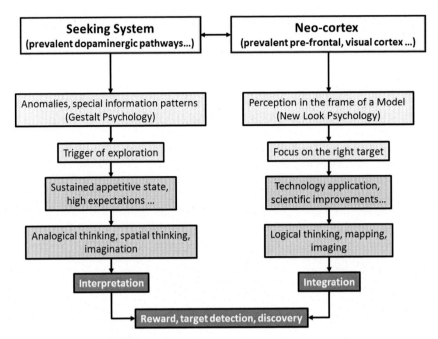

FIGURE 3.2   The structure of the "seeking process."

internal conditions (homeostasis[29]). In geosciences, the target is generally buried in the Earth interior. It means that exploration geoscientists search for something that is not visible at surface. However, the target itself can produce observable and measurable surficial effects[30] that trigger the exploration activity. This was the case, for instance, of the seepages of Tramutola. Besides the clear evidence of oil/gas seeps, geologists and geophysicists look for many other types of data at surface, such as geophysical anomalies, unusual natural phenomena, stratigraphy patterns, spatial distribution of outcrops, and their mutual correlations.

The effectiveness of surficial observations in triggering exploration depends on geological knowledge. In fact, exploration geophysicists and geologists are used to "interpret" anomalies and patterns of information

---

[29] Homeostasis is any self-regulating process by which biological systems tend to maintain stability while adjusting to conditions that are optimal for survival (Encyclopedia Britannica). For instance, energy balance is the homeostasis of energy in living systems. In case of negative balance, the animal starts seeking for food.

[30] For instance, geophysicists are able to measure at surface, or in boreholes, the Earth response to naturally or artificially induced physical fields (seismic, electromagnetic, gravity, magnetic).

in the frame of a model. The exploration work is successful especially when we know the context of our research. For that reason, exploring the sub-soil with a robust geological model in mind is much more effective than exploring just based on some type of anomaly.

3.5.2 Sustaining the Process

After the seeking (exploration) process has been triggered, technology, knowledge, science, and many types of resources are necessary for obtaining (discovering) the final reward (target). The activation of the DA pathways allows a *sustained appetitive state*, strong motivation, enthusiasm for novelties, and high expectations. These are the key emotions stimulating analogical and spatial thinking and, over all, imagination. In exploration geosciences, the rational counterparts of spatial and imaginative attitudes are represented by mapping and imaging technology.

3.5.3 Driving the Process to Success

Exploration geosciences are strongly based on interpretation of information. This is often characterized by high degree of uncertainty and ambiguity (Dell'Aversana, 2013). To perform a successful exploration workflow, geoscientists must be able to solve a sort of jigsaw puzzle. They must create coherent geological models from heterogeneous data sets (multiscale, multiphysics, multidimensional information). Human brain evolved over 2 million years and is perfectly structured for *integrating complex information* streams. Indeed, integration is the main factor driving the seeking process toward the success. Technologies and workflows should be in line with that intrinsic integration ability of our brain.

These remarks, supported by the scheme of Fig. 3.2, indicate the road map for developing new technology in exploration geosciences, consistently with the key features of our brain. This is what I call "brain-based technologies" and I will discuss extensively about it in the third part of this book. Multimodal imaging, multisensory perception, pattern recognition, and full cognitive integration represent the principal drivers of this geocognitive approach.

References

Alcaro, A., Huber, R., Panksepp, J., December 2007. Behavioral functions of the mesolimbic dopaminergic system: an affective neuroethological perspective. Brain Res. Rev. 56 (2), 283–321. http://dx.doi.org/10.1016/j.brainresrev.2007.07.014. Published online 2007 Aug 21.

Arias-Carrión, O., Stamelou, M., Murillo-Rodríguez, E., Menéndez-González, M., Ernst Pöppel, E., 2010. Dopaminergic reward system: a short integrative review. Int. Arch.

Med. 3, 24. http://dx.doi.org/10.1186/1755-7682-3-24. http://www.biomedcentral. com/1755-7682/3/24/.

Baldacci, L., Viola, C., 1894. Sull'estensione del Trias in Basilicata e sulla tettonica generale dell'Appennino meridionale. Boll. Com. Geol. D'It. 25 (4), 372–390. Roma.

Blood, A.J., Zatorre, R.J., 2001. Intensely pleasurable responses to music correlate with activity in brain regions implicated in reward and emotion. Proc. Natl. Acad. Sci. 98, 11818–11823.

Caire, A., Glangeaud, L., Grandjaquet, C., 1960. Les grand traits structureaux et l'évolution de territoire calabro-sicilien (Italie méridionale). Bull. Soc. Geol. Fr. Ser. 7 2, 915–938.

Centonze, D., Picconi, B., Gubellini, P., Bernardi, G., Calabresi, P., 2001. Dopaminergic control of synaptic plasticity in the dorsal striatum. Eur. J. Neurosci. 13 (6), 1071–1077.

Crema, C., 1902. Il petrolio nel territorio di Tramutola (Potenza). Boll. Soc. Geol. Ital. 21, 36–38. Roma, It.

Damasio, A., 2010. Self Comes to Mind: Constructing the Conscious Brain. Pantheon, New York.

Dell'Aversana, P., June 2001. Integration of seismic, magnetotelluric and gravity data in a thrust belt interpretation. First Break 19 (6), 334–335. http://dx.doi.org/10.1046/j.1365-2397.2001.00158.x.

Dell'Aversana, P., November 2003. Integration loop of 'global offset' seismic, continuous profiling magnetotelluric and gravity data. First Break 21, 32–41.

Dell'Aversana, P., 2014. Integrated Geophysical Models: Combining Rock Physics with Seismic, Electromagnetic and Gravity Data. EAGE Publications.

Dell'Aversana, P., 2013. Cognition in Geosciences: the Feeding Loop Between Geodisciplines, Cognitive Sciences and Epistemology. EAGE Publications, Elsevier.

Esestime, P., Hewitt, A., Hidgson, N., February 2016. Zohr – A newborn carbonate play in the Levantine Basin, East-Mediterranean. First Break 34, 87–93.

Flores, G., 1955. Discussion. In: Beneo, E. (Ed.), Les resultats des études pour la recherche petrolifère en Sicile. 4th World Petroleum Congr., Rome, 1955, Proceedings, Sect. 1/A/2, 19–21.

Ginzburg, C., Davin, A., Spring 1980. Morelli, Freud and Sherlock Holmes: clues and scientific method. In: History Workshop, No. 9. Oxford University Press, pp. 5–36. Stable URL: http://www.jstor.org/stable/4288283.

Glangeaud, L., 1952. Interprétation téctonophysique des caractères structureaux et paléogéographiques de la Méditerranée occidentale. Bull. Soc. Géol. Fr. Ser. 6 5 (6), 867–891.

Greengard, P., 2001. The neurobiology of dopamine signalling. Biosci. Rep. 21 (3), 247–269.

Grandjacquet, C., 1961. Aperçu morphotectonique et paléogéographique du domaine calabro-lucanien (Italie méridionale). Bull. Soc. Géol. Fr. Ser. 7 3, 610–618.

Huang, Y.Y., Simpson, E., Kellendonk, C., Kandel, E.R., 2004. Genetic evidence for the bidirectional modulation of synaptic plasticity in the prefrontal cortex by D1 receptors. Proc. Natl. Acad. Sci. U S A. 101 (9), 3236–3241.

Kastens, K.A., Manduca, C.A. (Eds.), 2012. Earth and Mind II: A Synthesis of Research on Thinking and Learning in the Geosciences. The Gological Society Of America, Special Paper 486.

Kastens, K.A., Ishikawa, T., 2006. Spatial thinking in the geosciences and cognitive sciences: a cross-disciplinary look at the intersection of the two fields. In: Manduca, C.A., Mogk, D.W. (Eds.), Earth and Mind: How Geologists Think and Learn about the Earth, pp. 53–76. http://dx.doi.org/10.1130/2006.2413(05). Geological Society of America Special Paper 413.

Kuhn, T.S., 1962. The Structure of Scientific Revolutions. University of Chicago Press, Chicago, USA.

Li, S., Cullen, W.K., Anwyl, R., Rowan, M.J., 2003. Dopamine-dependent facilitation of LTP induction in hippocampal CA1 by exposure to spatial novelty. Nat. Neurosci. 6 (5), 526–531.

Liu, Y.C., Sachs, B.D., Salamone, J.D., 1998. Sexual behavior in male rats after radiofrequency or dopamine-depleting lesions in nucleus accumbens. Pharmacol. Biochem. Behav. 60 (2), 585–592.

Merla, G., 1952. Ricerche tettoniche nell'Appennino settentrionale. In: Int. Geol. Congr., Rep. 18th Session. Great Britain. Proc. Part XIII, London. Part III acoording to Fabiani & Segre (1952).

Migliorini, C., 1949. I cunei composti nell'orogenesi. Boll. Soc. Geol. Ital. 67, 29–142.

Mostardini, F., Merlini, S., 1986. Appennino centro meridionale Sezioni geologiche e proposta di modello strutturale. Mem. Soc. Geol. It. 35, 177–202.

Nestler, E.J., 2004. Molecular mechanisms of drug addiction. Neuropharmacology 47 (1), 24–32.

Panksepp, J., 1998. Affective Neuroscience: The Foundations of Human and Animal Emotions. Oxford University Press, Oxford.

Panksepp, J., Biven, L., 2012. The Archaeology of Mind: Neuroevolutionary Origins of Human Emotions (Norton Series on Interpersonal Neurobiology).

Paulsen, O., Sejnowski, T.J., 2000. Natural patterns of activity and long-term synaptic plasticity. Curr. Opin. Neurobiol. 10, 172–179.

Reynolds, S.J., 2012. Some Important Aspects of Spatial Cognition in Field Geology. The Geological Society of America, Special Paper 486, pp. 75–77.

Scandone, P., Sgrosso, I., 1964. Flysch con Inocerami nella valle del Cavolo presso Tramutola, (Lucania). Boll. Soc. Natur. Napoli 166–175, 1 tav., Napoli.

Scandone, P., 1967. Studi di geologia lucana: Carta dei terreni della serie calcareo-silico-marnosa e i suoi rapporti con l'Appennino calcareo. Boll. Soc. Natur. Napoli 76, 301–469.

Scandone, P., Bonardi, G., 1967. Synsedimentary tectonics controlling deposition of Mesozoic and Tertiary carbonatic sequences of areas surrounding Vallo di Diano (Southern Apennines). Mem. Soc. Geol. It. 7, 1–10.

Scandone, P., 1968. Sul significato dei "calcari con liste e noduli di selce" di S. Fele e delle brecciole calcaree negl scisti silicei della Lucania. Boll. della Soc. dei Nat. Napoli LXXVI (Parte Prima), 189–197.

Scandone, P., 1969. Outlines of the Geology of the Lucanian Apennines. Guida per una escursione nell'Appennino campano-lucano, 21 pp., Napoli.

Scandone, P., 1971. Note illustrative della Carta Geologica d'Italia alla scala 1:100000. Fogli 199 e 210. "Potenza" e "Lauria". Nuova technica grafica, Roma.

Scandone, P., 1972. Studi di geologia lucana: Carta dei terreni della serie calcareo-silico-marnosa e note illustrative. Boll. Soc. Natur. Napoli 81, 225–300.

Scandone, P., 1975. The preorogenic history of the Lagonegro Basin (Southern Apennines). In: Squyres, C. (Ed.), Geology of Italy, The Earth Sciences Society of the Libyan Arab Republic. Tripoli.

Sclater, J.G., Anderson, R.N., Bell, M.L., 1971. Elevation of ridges and evolution of the central eastern Pacific. J. Geophys. Res. 76, 7888–7915.

Signorini, R., 1939. Sulla tettonica dei terreni mesozoici nell'Appennino lucano. Rend. Acc. Naz. Lincei. Sc. Fis. Ser. 6a 29, 558–562.

Van Dijk, J.P., Affinito, V., Atena, R., Caputi, A., Cestari, A., D'Elia, S., Giancipoli, N., Lanzellotti, M., Lazzari, M., Oriolo, N., Picone, S., 2013. Cento anni di ricerca petrolifera. In: Atti del Primo Congresso de Geologi di Basilicata - Ricerca, sviluppo ed utilizzo delle fonti fossili - Il ruolo del geologo, pp. 29–77.

Wolf, M.E., Mangiavacchi, S., Sun, X., 2003. Mechanisms by which dopamine receptors may influence synaptic plasticity. Ann. N. Y. Acad. Sci. 1003, 241–249.

CHAPTER

4

Imaging

4.1 GEOPHYSICAL PROBLEMS

I would like to use again Fig. 1.1 discussed in Chapter 1 for introducing one of the key aspects of geophysics and exploration geosciences: imaging. That figure shows a *geophysical image* of the subsoil, down to depth of a few meters below the ground of an ancient church. Using that image, I was able to identify my exploration target, map it in the space, define its geometrical properties (size, depth, etc.), and detect the exact location of important details (such as the trapdoor). Having a good image of the target was a crucial requisite for completing successfully my archaeological investigation and accessing into the underground room.

Images represent the fundamental *bricks* in geosciences for *building* Earth models and taking decisions, such as *"if* and *where* to drill a well."* For analogous reasons, imaging is a fundamental process for the study of human body for clinical analysis, medical diagnosis, and intervention. In medicine as well as in geosciences, significant efforts, in terms of research and budget allocation, are addressed to develop imaging technologies. Medical imaging includes X-ray radiography, magnetic resonance imaging (MRI), medical ultrasonography or ultrasound, endoscopy, thermography, medical photography, positron emission tomography (PET), single-photon emission computed tomography (SPECT), and other advanced technologies. Geoscientists use many different types of geophysical imaging techniques, based on different physical principles and addressed to different exploration purposes. For instance, in hydrocarbon exploration, the dominant approach is based on seismic imaging including seismic data migration methods, transmission tomography, and so forth. Some examples of seismic images are provided in Chapter 2 (Figs. 2.5–2.7). An extensive discussion about geophysical data analysis, processing, and imaging is provided by Yilmaz (2001). Different types of imaging in exploration geophysics are obtained using electrical potentials and electromagnetic fields (Zhdanov, 2009). Well-known examples are electrical resistivity tomography (ERT), electromagnetic controlled source data inversion, and magnetotelluric data inversion (Dell'Aversana, 2014a).

The final goal of geophysical imaging is to provide the geoscientists with a high-resolution image of a portion of the subsoil. Complex processing workflows are addressed to accomplish that objective. Unfortunately, the path toward a good image of the subsoil is often very rough. In fact, the physical fields (seismic, electromagnetic, magnetic) bringing information to surface about the deep exploration objective, arrive to the recording sensors after strong attenuation and scattering effects. Consequently, the final image of the subsoil obtained through analysis of geophysical signals recorded at surface can be unclear, ambiguous, and affected by artifacts. Moreover, different from imaging techniques used in medicine, in geophysics both sources and recording devices are commonly deployed at surface[1] (excluding the cases of borehole, airborne, and satellite remote sensing methodologies). This type of acquisition geometry is not optimal for obtaining a good *coverage*

---

[1] This surface can be the sea level, land topography, or marine bathymetry.

(sampling) of the investigated portion of the Earth, especially in case of tomography. The undesired effect is that the subsoil is often "under-sampled" by the physical waves used in geophysical prospecting. Consequently, the final Earth models can be affected by significant un-certainties and by poor signal-to-noise ratio. All these problems are increased in case of complex geological settings, where strong effects of scattering and dispersion of wave fields can predominate on useful signals.[2] All these geophysical problems are well known in the practice of exploration geosciences. As I anticipated in Chapter 2, integration of multidisciplinary geophysical data represents the ideal approach for minimizing the limitations and maximizing the benefits of each individ-ual method. In fact, an extended scientific literature demonstrates that seismic, electromagnetic, gravity, and magnetic methods are effective especially when they are properly combined into integrated workflows. There is an equivalent "principle of integration" in medicine, where combination of complementary methods of analysis is mandatory for obtaining a reliable diagnosis and/or an accurate investigation of the body interior. Indeed, neuroimaging or brain imaging combines a multitude of techniques to directly or indirectly image the structure and the main functions of the nervous system. In medical sciences as well as in exploration geosciences, the final objective is to obtain a robust model of the investigated target; the integrated imaging process is determinant for reaching that result.

Before discussing the technical aspects of that process, several funda-mental questions must be investigated. Why images are so important for our reasoning and for creating a model of a "piece of reality"? What is the neurobiological background of imaging and, in particular, of geophysical imaging? Is imaging just a visual representation of the reality or some-thing more? Can we create "the image of a sound"? What is multisensory imaging? What is the relationship between imaging and mapping? What is a mental image? What is a mental map? What is the difference between imaging, imagery, and imagination? What are the links between these concepts? Can we enhance our interpretation and our technology in geosciences improving our understanding of the neurobiological back-ground of imaging, imagery, and imagination?

In this chapter, I will attempt a preliminary analysis of the above open questions starting from the original problem: mental imagery. This is an ancient question involving fundamental cognitive aspects, many of which strictly related with exploration geosciences. The following analysis is useful for creating the background of a new generation of brain-based

---

[2] An extensive discussion about these imaging problems in complex geological settings is provided in Dell'Aversana (2014a).

technologies driven by the concepts of *optimized visual cognition, multi-sensory imaging*, and *expanded integration*.

4.2 MENTAL IMAGERY

Our thought is largely based on "mental imagery." Following the Stanford Encyclopedia of Philosophy, "Mental imagery (varieties of which are sometimes colloquially referred to as *visualizing, seeing in the mind's eye, hearing in the head, imagining the feel of* etc.) is quasi-perceptual experience; it resembles perceptual experience, but occurs in the absence of the appropriate external stimuli. It is also generally understood to function as a form of *mental representation*."[3]

It is difficult to explain high-level cognition without making appeal to the storage and processing of *mental representations* in the form of images. Consider, for example, a familiar situation: imagine that you are going to work using your car, and imagine driving along the road from your house to your office. You can "see" in your mind a multitude of familiar objects, such as buildings, shops, and so forth. These represent a sort of complex mental picture produced by your mind. The view that considers the thought as formed by a sequence of mental pictures, or mental images, is traditionally known as "pictorialisms." That basic concept is very old. It was grasped by the ancient Greek philosopher Aristotle (384–322 BC). He supposed that mental images play an essential role in human cognition because they allow the process of abstraction and conceptualization. It is not difficult to agree with that basic idea because "mental imagery" is a familiar experience. However, the precise meaning of "mental image" and its neurobiological background has been difficult to define for long time. The ambiguity of this concept generated confusion and controversial among philosophers, psychologists, and cognitive scientists. Some of them such as Dennett (1969, 1979) and Pylyshyn (1973, 1978) advocate an alternative view called "descriptionalisms." While pictorialists think that mental images *represent* roughly the thought in a pictorial way, descriptionalists claim that natural language descriptions constitute the basic *representation* of our thought. For instance, I can describe the path from my house to the office using words and statements. Every mental image can

---

[3] Mental imagery is often used as a synonym of *imagination*. These concepts are obviously related, but they do not indicate necessarily the same thing. In fact, imagery is referred to a common ability of people, including those with scarce imagination. We will see that creating mental images depends on the structure and the topology of human brain, including the brain of people with scarce fantasy. Instead, *imagination* is sometimes related to the ability to think with fantasy. In this book, I intend *imagination* as a *creative attitude to mental imagery*.

be represented using the natural language, and this is what description-alists assume. The dispute between pictorialists and descriptionalists is known as the *imagery debate*. The debate is not over whether we "think in images" or "think in words." Both pictorialists and descriptionalists accept the idea that we can think through a sequence of images. The debated question is how our mind *represents* images in mind: pictorially or verbally. This is a "slippery" discussion. Thus, we need to start from the beginning of the story.

4.2.1 The Origins

Mental imagery implies the concept of *ideas* and *mental representation*. These concepts have interested philosophers for many centuries, from Plato (428/427 or 424/423–348/347 BC) to Descartes (1596–1650); the debate still involves modern neurobiologists and cognitive scientists such as Damasio, Edelman, Freeman, Hofstadter, Dennett, and many others. The crucial questions are related to the link between ideas, mental images, language, neurobiological background of images, and connections to objects belonging to the "internal and external world."

Although Plato does not discuss mental imagery systematically, in the *Theaetetus*[4] he argues that memory might be analogous to a wax tablet into which our perceptions and thoughts are stamped like images of them-selves. The concepts of *idea* and *image* in Plato are strictly related to the process by which human beings develop their knowledge about the world. The ancient philosopher makes a substantial distinction. In the *Timaeus*[5] he states that the robustness of knowledge depends on the realm from which it is obtained. In fact, the knowledge obtained by the flux of experience, which is continuously changing, will correspond to mere contingent opinions. Instead, the knowledge derived from nonempirical ideas is necessary and stable.

The Theory of Forms (Greek: ἰδέαι) typically refers to the belief expressed by Socrates in some of Plato's dialogs: the material world as it seems to us is not the real world; instead it represents only an *image* or *copy* of the real world. Thus, if we try to derive our knowledge from the experience, it will consist of mere opinions because the world of senses is in flux. This type of knowledge will be characterized by a lack of necessity and stability. On the other hand, if we derive our knowledge from uni-versal forms (the ideas), it will be necessarily an objective form of knowledge.

[4] The Theaetetus is one of Plato's dialogs concerning the nature of knowledge. It was written c. 369 BC.

[5] The Timaeus is another Plato's dialog, mostly in the form of a long monologue, written c. 360 BC.

It is well known that both Aristotle and Plato had a strong impact on the Western philosophy. Neoplatonism was the dominant philosophical system of later antiquity, influencing late Roman and early medieval Christian thought. It was not just a revival of Plato's ideas but was based mainly on Plotinus' synthesis. This included the teachings of Plato, Aristotle, Pythagoras, and other Greek philosophers. Plotinus had extremely modern ideas about imagery and imagination. His view was close to contemporary theories about perception and imagery, regarding both perceiving and imagining to be forms of *active receptivity*[6] toward the object of the experience.

Neoplatonism continued to influence the philosophical thought during the whole Renaissance, combining the ideas of Christianity with a new interpretation of the writings of Plato and Aristotle. The Italian Renaissance philosopher Gianfrancesco Pico della Mirandola, in around 1500, published a work entitled "De Imaginatione" (On the Imagination). That work was strongly based on Aristotle's concepts about mental imagery, such as *phantasia* and *phantasmata*.

René Descartes (1596–1650) describes his point of view about the relationships between the external world and our mental images in the *Treatise of Man*, where he discusses his physiological theory of visual perception. The philosopher states that the surface of the pineal gland is the "seat of imagination" and that "the images traced there are *ideas*" [idées] (Descartes 1664, p. 86 in Hall's translation).

One of Descartes' followers, Louis de La Forge (1632–66), suggested that the term "idée" should be applied only to concepts in the intellect. He invented the expression of "espèces corporelles (corporeal species)" to refer to the attitude of human imagination to create *pictorial images*.

Mental images are central to the cognitive theory of Thomas Hobbes (1588–1679). He did not make any clear distinction between images formed in the brain and ideas in the mind, using the term "idea" as a synonym for "image." That was an intuition not far from the modern neurobiological concept of mental image (Damasio, 2010), which is strictly related to the formation of neural aggregates (or neural maps) in the cerebral cortex. According to Hobbes, images are processes rather than entities. Our thoughts are associatively connected sequences of images passing through the mind.

Locke (1632–1704) referred to ideas as "the pictures drawn in our minds" and suggested an analogy between mental images and optical images (Essay II.xi.17).

---

[6] Following the paradigm of the "New Look Psychology," perception is not a passive cognitive state, but it is influenced by our prior knowledge, expectations, and cultural factors.

Kant (mainly in the period 1781–87) argued that ideas represent a sort of *vehicles of thought*: images play a significant role for connecting concepts to empirical reality. The *imagination* (einbildungskraft) allows us synthesizing the extreme complexity of our sensory perception into a coherent, meaningful image.

4.2.2 Toward a Modern Concept of Imagery

In the late 19th century, psychology began to emerge as an experimental science. Despite the impossibility to demonstrate the existence of mental images, the central role of imagery in mental life was not in question. However, at the beginning of the 20th century, experimental psychologists (especially those following the so-called behaviorist's approach) began to become skeptical about the whole notion of mental imagery. Of course, the physiological debate had a deep impact on philosophy. From one side, philosophers such as Russell, Bergson, and Frege still gave imagery a fundamental role in cognition and epistemology, and, in particular, in the theory of *meaning*. Instead, Wittgenstein rejected the hypothesis that the *meaning* could be based on imagery. Especially in the posthumously published writings, he expressed his skeptical position about the concept of the image as a sort of inner picture. Moreover, he rejected the empiricist view that thinking is based on images, regarding language as the predominant vehicle of thought. Other philosophers after Wittgenstein, such as Harrison, Goodman, and Fodor, agreed about the irrelevance of imagery to semantics. One of the main arguments used for supporting the cognitive unimportance of imagery was the following. Mental images are possible only when they are applied to concrete nouns; instead, many other linguistically expressible concepts, such as logical relations, cannot be translated into any image. For instance, it is easy to imagine a house, but it is difficult to create any image for words such as "not," "or," "if," and so forth.

After the predominant skeptical position of many psychologists and philosophers during the period of the behaviorist intellectual hegemony (1920–50), a revival of interest in imagery was a central aspect in the years of the *cognitive revolution* in psychology (during the 1960s and early 1970s). An important aspect that attracted the attention of modern psychologists was the central role of imagery in memory. Other experimental facts were determinant for that "imagery revival," such as the results of researches on hallucinogenic drugs, the confirmation of mnemonic effects of imagery, the developments in electroencephalography (EEG), and the finding that direct stimulation of specific areas of the brain can produce vivid imagery effects. The foundation of specialized journals and associations (Journal of Mental Imagery; Imagination, Cognition, and

Personality; Association for the Study of Mental Imagery; International Imagery Association) confirmed the increasing interest about imagery. This was considered as a fundamental aspect of human cognition.

In the early 1970s, Roger Shepard and his research team demonstrated the "mental rotations of images" and the attitude of some people to perform "mental scanning of visual images." These scientists stated that mental imagery has inherently spatial properties. In other words, they suggested a strict connection between imagery and spatial thinking. These connections cannot be found in any type of symbolic representation, including language.

The neuroscientist W. J. Freeman has recently introduced many original ideas expanding the concept of imagery, showing that it is related to a wide range of high-level cognitive functions. For instance, he introduced the concept of "olfactory imagery" as the result of special neural dynamics observed in the brain of some animals (rabbits) during laboratory experiments. Moreover, Freeman suggested the idea of "search image": this is a form of selective attention of many predator species aimed at recognizing specific preys in their environment. The work of this neuroscientist is extremely important and particularly useful for introducing the neurobiological aspects of mental imagery.

4.3 NEUROBIOLOGICAL BACKGROUND

What happens in the brain when we "see images with the mind's eye"? Is there any neurobiological explanation of imagery? Can we describe a mental image in terms of neural activity? In the previous paragraphs, we have seen that scarce or null experimental evidences supported the first pioneering hypothesis of cognition in the form of images and mental representations. Unfortunately, the lack of direct observations represented the main limitation of all the theories of cognition based on *imagery*. A radical change happened in the second half of the 19th century. In 1875, Richard Caton (1842−1926) demonstrated the possibility of electrical phenomena of exposed cerebral hemispheres of rabbits and monkeys. A few years later, Adolph Beck published his studies about spontaneous electrical activity of animals, measured through electrodes placed directly on the surface of the brain during sensory stimulation. These tests provided the first experimental base of a noninvasive approach for "observing" the brain at work: EEG. In this technique, electrodes are placed on the scalp to measure voltage fluctuations caused by the brain electrical activity. The analysis of an EEG is commonly focused on the type of neural oscillations and their frequency content observed in EEG signals. Many diagnostic applications are possible with this approach, including epilepsy. It is remarkable that the diagnosis is generally based on the evidence of *anomalies* in EEG readings, as in many geophysical investigations. EEG represents just one example among many other noninvasive techniques used for "imaging" the activity of the brain. Nowadays, modern technologies of neuroimaging allow investigating the relationship between the activity of specific areas of the brain and mental functions. For instance, recent studies on blind individuals have demonstrated that the performance of nonvisual tasks, such as Braille reading, and various auditory functions can activate posterior visual areas. This is a clear indication that the formation of images in our brain is not exclusively confined to the visual sense. Imagery is a general cognitive activity of our brain. Indeed, mental images are strictly connected to the so-called *brain maps*. These represent large neural populations activated in a synchronous way, "… as the result of the momentary activity of some neurons and of the inactivity of others" (Damasio, 2010). As shown by Freeman, brain maps and neural oscillations represent crucial aspects of the dynamic organization of the brain and, consequently, of high-level cognition.

4.3.1 From Chaos to Multisensory Brain Maps

Walter Jackson Freeman conducted pioneering studies on how the brain produces meaning (Freeman, 1987, 1999, 2000, 2007; Freeman and Vitiello, 2006; Freeman et al., 2000). By using recent results in brain imaging and theories of chaos and nonlinear dynamics, Freeman shows how the brains create intention and meaning. He starts from the concept that neural populations show all the main characteristics that define the open-complex systems. In fact, they consist of many independent elements, the neurons. These are reciprocally connected and form large populations. Moreover, the link between two neurons is not crucial for defining the behavior of the neural aggregate, and the input—output relationship for each neuron is nonlinear. As it happens in all the complex systems, also in the brain, collective action of a neural population cannot be described as the linear sum of the single individual neural activities. When the network of dendrites and axons is sufficiently complex, the group of connected neurons stops to be an aggregate and becomes a *population*. This is a real transition of state. Every neural population is characterized by a stationary level of activity in terms of global oscillation, independently from the strength of the single connections. This represents the background activity of the cortex. If a neural population is perturbed by an external impulse, such as an electrical stimulus, the level of activity increases for a while but then it gradually returns to its stationary state. The activity consists of oscillations of the electrical potential. This can be measured using electrodes properly located in the cerebral cortex.

Freeman argues that the populations have a fixed point attractor[7] because after the effect of the external stimulus they return to the same initial level of activity. The totality of the amplitude values of the neural oscillations form the space of the states of the population. The part of the space corresponding to the stationary state of neural population represents a basin of attraction. Normally the response to an impulse is oscillatory, with frequency variable between 20 and 100 Hz. In presence of a significant external stimulus, the oscillation of a neural population can grow and does not return to its original attractor. In fact, many neurons can interact constructively producing a positive feedback until the neural population reaches a new stable oscillatory state different from the original. This is a new attractor because it tends to become stable and is considered by Freeman such as a limit cycle.

Following Freeman, the space of states of the cortex is formed by a landscape of attractors with many contiguous basins of attraction. These correspond with dynamic configurations of neural activity in response to external stimuli. In this way, the experience forms progressively the space of states of the neural populations in the cortex. In summary, Freeman significances and concepts correspond with basins of attractions in the space of states defined by the properties of oscillations of neural populations. Fig. 4.1 shows instructive examples of maps of electrical potentials illustrating how the brain works through maps formed by some physical properties (such as electrical potential) of neural populations oscillating synchronically (in the same time frame, with the same frequency).

The activation of neural maps and, consequently, the formation of mental images are made possible by the structure of the cerebral cortex (Damasio, 2010).[8] This is characterized by a layered structure. There are six main layers, containing characteristic distributions of neuronal cell types and connections with other cortical and subcortical regions. Each cortical layer has a sheath-like structure resembling a two-dimensional (2D) square grid. The following words of Damasio clarify perfectly how this type of topology allows the formation of neural patterns, efficient connections, and, finally, brain maps. "The main elements in the grid are neurons, displayed horizontally. You can imagine something like the plan of Manhattan, but you must leave Broadway out because there are no major oblique lines in the cortical grids. The arrangement, you immediately realize, is ideal for overt topographical representation of objects and

---

[7] An attractor is a set of numerical values toward which a system tends to evolve, starting from a wide variety of initial conditions.

[8] Cerebral cortex is the part of the brain where most frequently neural maps are formed, although these can be formed also in subcortical structures, such as certain types of nuclei.

FIGURE 4.1   Examples of maps of neural electric potential measured through electroencephalogram during experiments on the cortex of rabbits. *After Freeman, W.J., 2000. Neurodynamics. An Exploration in Mesoscopic Brain Dynamics, Springer, London.*

actions. Looking at a patch of cerebral cortex, it is easy to see why the most detailed maps the brain makes arise here, although other parts of the brain can also make them, albeit with a lower resolution" (Damasio, 2010).

Neural maps have two fundamental characteristics. First, they continually change, reflecting the events happening in the interior of our body and our interactions with the environment around us. When we move through the world, adapting our behavior to the continuous external changes, or when we move our body to reach some type of objective, neural maps also change. Furthermore, maps' dynamics reflect in some way the changes of our emotions and feelings.

Second, the maps are continuously combined and recombined with other maps, through stacking of different cortical layers. Dynamic integration of neural maps is a diffuse feature of the brain. The neurons of stacked cortical layers form vertical arrays of elements called "columns" including hundreds of neurons. These columns represent the pathways of input—output information between different cortex layers. Furthermore, they allow communication among the cortex itself, the sensory system, and the rest of the body.

The cerebral cortex is not the only place in the brain where neural maps are formed. Several subcortical structures, such as the geniculate bodies, the colliculi, the nucleus tractus solitarius, and the parabrachial nucleus, can create maps, although with lower resolution than the cortical maps. A fundamental work is performed by the superior colliculi (see Fig. 2.2 in Chapter 2). They are formed by seven layers: the superficial layers have to do with vision, for instance, the layer II receives signals from retina and from the primary visual cortex. Other surficial layers assemble the retinotopic map of the contralateral visual field. The deep layers contain also maps of auditory and somatic information. A crucial point is that all these maps, visual, auditory, and somatic, are spatially well correlated. They are stacked, in such a way that multisensory perception can be easily integrated. In other words, the topologic organization of the various types of maps in the colliculi is extremely suitable for facilitating the combination of different types of sensory stimuli. Thus, *the brain is not only a good cartographer,* as stated by Damasio, but also integrates the maps that it produces in response to different perceptive modalities. That integration attitude of the brain offers obvious adaptive benefits.[9]

There is a generic correspondence between mapped neural patterns and the actual external objects. For instance, experiments on monkeys demonstrated that there is correlation between the structure of a visual stimulus and the neural activity it evokes. Also in humans, certain patterns of activity in sensory cortices show a good topological correspondence to a certain class of objects. Although that is just a weak correspondence (we cannot observe in terms of neural activations the same images a monkey or a man sees), it helps to clarify from a neurobiological point of view the crucial questions of imagery: "what is a mental image and how our brain creates it?"

Recently, Huth et al. (2016) showed that the meaning of language is represented in regions of the cerebral cortex through the activation of neural maps. These areas of the brain form the so-called "semantic system."[10] The authors mapped the semantic activity of the brain of subjects by functional MRI while listening to narrative stories. These maps show that "the semantic system is organized into intricate patterns that seem to be consistent across individuals." This interesting work shows that also abstract terms of the language are represented in the brain in terms of neural maps. This is a strong evidence supporting the pictorial nature of our mental representations.

Indeed, following Huth, Damasio, Freeman, and many other cognitive scientists, the pictorialist interpretation seems to be more appropriate than the descriptionalist view. In fact, mental images are not just verbal representations; they have a neurobiological/physical nature. They derive from the continuous activity of cortical and subcortical structures of the brain to form neural maps from multisensory perception. Using the words of Damasio, "The mapped patterns constitute what we, conscious creatures, have come to know as sights, sounds, touches, smells, tastes, pains, pleasures, and the like; in brief, images. The images in our minds

---

[9] In this book, I will discuss how that capability of integration can be used for creating advanced technology for signal processing and interpretation in geophysics.

[10] I remark that "semantic system" represents a central concept developed by Dell'Aversana (2013) for describing the cognitive processes of significant construction. The work of Huth et al. provides imaging evidences of the activity of our brain while it works as a semantic system.

are the brain's momentary maps of everything and of anything, inside our body and around it, concrete as well as abstract, actual or previously recorded in memory. The words I am using to bring these ideas to you were first formed, however briefly and sketchily, as auditory, visual, or somatosensory images of phonemes and morphemes, before I implemented them on the page in their written version. Likewise, the written words, now printed before your eyes, are first processed by you as verbal images (visual images of written language) before their action on the brain promotes the evocation of yet other images, of a nonverbal kind. The nonverbal kinds of images are those that help you display mentally the concepts that correspond to words. The feelings that make up the background of each mental instant and that largely signify aspects of the body state are images as well. Perception, in whatever sensory modality, is the result of the brain's cartographic skill … In conclusion, images are based on changes that occur in the body and brain during the physical interaction of an object with the body" (Damasio, 2010).

4.3.2 From External Images to Mental Images

In the previous paragraph, we have seen that the human brain continuously creates mental images of the external/internal world in terms of dynamic neural maps. That is a neurobiological explanation of the fact that imaging techniques play such an important role in geology, geophysics, and all the other Earth disciplines. Images are fundamental elements of our cognition, particularly in our field of study. In fact, geoscientists produce maps and images of many different types and for a wide range of purposes: topographic and geological maps, seismic images of the subsoil, three-dimensional (3D) cubes of some geophysical attribute, and so on. In this paragraph, I would like to summarize, in a schematic way, how external images are translated into mental images.

Modern neurosciences teach that the root of the mapping/imaging workflow is our cerebral cortex in cooperation with several subcortical nuclei. In fact, our cortex and the colliculi are structured for producing mental maps and images and for integrating them.

Fig. 4.2 summarizes the process that transforms the "traditional" geoimages (maps, sections, volume attributes, etc.) into the related mental images. I try to explain it schematically.

1. On the top left corner, a picture of Mt. Vesuvius represents a "piece of the external world." Here it is indicated with the R symbol.
2. Mapping and imaging the features of "a piece of the external world" represent crucial steps in the process of developing knowledge. Indeed, images are the *bricks* of human cognition. A geological map or a seismic section represents a sort of codification of the external

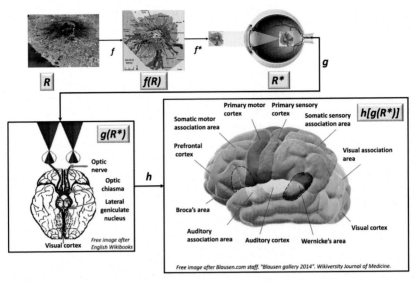

FIGURE 4.2   The cognitive path from the "external" world to the "internal" world. *After Dell'Aversana, P., July 2015. The brain behind the scenes — Neurobiological background of exploration geophysics, First Break 33, 41–47.*

world into the mind language. Here I indicate with the symbol f the process of creating a map of R (the Vesuvius map, in this example). This process includes historical reconstructions, work in the field, study of the outcrops, mineralogical analysis, etc. The map shown in the second panel on the top of Fig. 4.2 represents an isomorphic projection of R. It means that the spatial relationships are saved after the projection. The result is here indicated with $f(R)$. The next steps allow transforming the external image (the map of Vesuvius) into a mental image (the mental representation of the Vesuvius map) through basic perception and advanced cognitive processes.

3. Let us suppose that a volcanologist analyzes the Vesuvius map. This is perceived by the retinas, forming a retinoic image $f^*[f(R)] = R^*$. Here the symbol f^* represents a complex process that is well known by the experts in physiology of visual perception. The image formed on the retinas is a specular map of the original object.

4. Then, the retinoic image R^* is projected to the primary visual cortex (left bottom panel of Fig. 4.2). The retinotopy process (g) is the mapping of the visual input from the retinas to neurons within the visual stream. Although it is not explicitly showed in the figure, the retinotopic image $g(R^*)$ of the retinoic image R^* is a further

"quasiisomorphic" projection on the primary cortex. In fact, the spatial relationships of the original perceived object are partially saved after the projection from the retinas to the primary visual cortex.

5. Finally a mental map or a mental image[11] $h[g(R^*)]$ is created by further projections from primary cortex to high-order (secondary and associative) cortices. The associative cortices include the main part of the human cerebral surface. They are responsible for all those high-level functions that produce cognition and behavior (the processes by which we come to know the world and interact with it). All the primary cortices (e.g., visual, audio, motor cortex) cooperate with each other and with the high-order cortices (and with other parts of the brain) to produce the fundamental process of multimodal cognition. This advanced stage of human cognition involves the cooperation of all senses, memory, subcortical neural nuclei, etc.

In summary, the "real objects" (Vesuvius, in our example) are conceptually "transferred" from the physical world toward the brain through a sequence of projections of different order. The whole process is facilitated when some type of visual representation of the *external object* is created, such as a map or another type of image. This is the reason why mapping and imaging are so important in natural sciences such as the Earth disciplines. When geologists and geophysicists create a good image (a map, a section, a model) of the object under study (a geological region, a sedimentary basin, a hydrocarbon reservoir), the whole cognitive process of analysis and interpretation is improved.

A further comment: there is an important difference between external and internal images/maps. The first are projections of pieces of the world that can be shared (visually, symbolically, etc.) within a community of experts. Instead, "internal or mental images" are necessarily subjective and can be shared after they have been translated into some type of formal model or code, including language. The subjective nature of the mental images is one of the reasons why different people generally propose different interpretations of the same real object, as it frequently happens in the geophysical practice. An additional reason is that different people have different imaginations. I think that such unavoidable variety of interpretations can represent an advantage, rather than a limitation.

---

[11] I remark again that the concept of mental map or mental image is here intended as patterns of synchronized neural activity in sensory and associative cortices triggered by some external or internal stimuli.

4.4 LINKS AND EXAMPLES

4.4.1 Earth Model and Health Model

Because of their common neurobiological and cognitive background, it is reasonable to expect significant links and analogies between the imaging process in Earth and medical sciences. Investigating the relationships between these two fields of study can help to illuminate crucial aspects of both scientific domains. Moreover, ideas and hypotheses can be "exported" from one field to another. Indeed, despite the obvious differences related to technical details, spatial scale, and specific objectives, both types of imaging process share similar aspects and analogous purposes. A "picture" of the subsoil describing the spatial distribution of physical parameters contributes to create an *Earth model*. This is fundamental in exploration geosciences. Analogously, an image of the body interior contributes to define a diagnosis that is a sort of *model* about the health state of a patient.[12] Conceptually speaking, this *Health model* is not different from an *Earth model*. The first is based on medical criteria, such as analysis of symptoms; test results including blood pressure, checking the pulse rate, listening to the heart with a stethoscope, urine tests, blood tests, electrocardiogram, biopsy; and, of course, medical imaging. Instead, an *Earth model* is based on analysis of geological outcrops, borehole data, mineralogical data, paleogeography reconstructions, and geophysical imaging. Both the *Health model* and the *Earth model* are strongly[13] based on images, and both are crucial for driving the subsequent decisions and actions. These can be, for instance, a surgical intervention or drilling operations, respectively.

Fig. 4.3 shows an example of a 3D seismic velocity model (left panel) and a 3D child's brain imaging obtained using magnetic resonance (right panel). Different colors indicate variable values of seismic velocity (in the first case) and variable values of brain growth and loss of tissue (in the second case). The acquisition methods used in the two cases are different, but the imaging methods are conceptually similar: both apply *tomography inversion* (see Appendix 1) of some type of physical measurement for reconstructing the spatial distribution of a physical (or biological) parameter. In case of seismic tomography, the image is produced by geophysical inversion of seismic travel times. Instead, MRI, nuclear MRI, or magnetic resonance tomography is a medical imaging technique that uses magnetic fields, radio waves, and field gradients for imaging the

---

[12] Medical diagnosis can be defined as the process of *attempting to determine* the identity of a possible disease or disorder.

[13] ... But not exclusively.

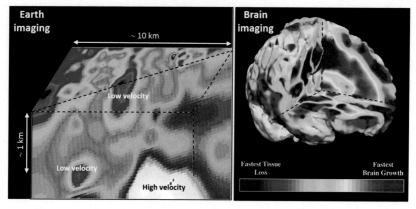

FIGURE 4.3   Left panel: 3D seismic velocity cube. Right panel: Brain imaging from magnetic resonance imaging scan data for child onset schizophrenia showing areas of brain growth and loss of tissue. *Credits: Image Library of National Institute of Mental Health, image in the public domain: https://images.nimh.nih.gov/public_il/searchresults.cfm.*

anatomy and the physiological processes of the body, such as the brain, in this particular example.

Another example of imaging in the geophysical domain is shown in Fig. 4.4. This is a resistivity section obtained by processing geoelectric data recorded during a geophysical experiment (Dell'Aversana, 2014a). ERT or electrical resistivity imaging is a geophysical technique used for imaging subsurface structures from measurements of electrical potential made at the surface or in borehole. Colors in Fig. 4.4 indicate different values of the electric resistivity in a basaltic sequence, down to a depth of few tens of meters below the surface. These basalts are not homogeneous and the colors allow immediate identification of resistivity variations associated to changes in lithology. There is a shallow part characterized by high resistivity (in red). A few meters below that there is a more conductive part (blue in the figure) corresponding to the presence of water in the most fractured portion (green) of the lava flow itself. Then, below the conductive interval, there is again a resistive electric response, corresponding to the massive (less fractured) part of the basalts.

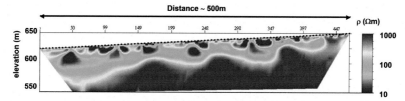

FIGURE 4.4   Resistivity section obtained by tomography of geoelectrical data. *After Dell'Aversana, P., 2014. Integrated Geophysical Models: Combining Rock Physics with Seismic, Electromagnetic and Gravity Data, EAGE Publications.*

What are the common features of the images shown in Figs. 4.3 and 4.4? What are the analogies in the process of medical and geophysical imaging? What are the main problems affecting both? What are the possible solutions?

4.4.2 Geophysical and Medical Imaging Based on Inversion

Despite the different finalities, applicative and technological aspects, medical and geophysical imaging share a similar mathematical background. In fact, they are largely (although not exclusively) based on numerical methods addressed to solve some type of *inverse problem* (Menke, 1989; Tarantola, 2005). Inversion means "using the result of some measurements to infer the values of the parameters that characterize the system under study" (see Appendix 1). For instance, geophysicists use electric potentials measured by electrodes deployed at surface to infer a subsurface model of electric resistivity (Nabighian, 2008; Dell'Aversana, 2014a). The physical principle used by geoelectric methodology is Ohm's law. It describes the flow of current in the ground following the well-known vector relation $J = \sigma E$, where σ is the conductivity of the medium (the reciprocal of the electric resistivity ρ), J and E are the current density and the electric field, respectively. Conductivity (or resistivity) measurements are made by injecting current into the ground through a set of *current electrodes* and measuring the resulting voltage difference between *potential electrodes*. The measured electric potential depends on the spatial distribution of the resistivity in the subsurface that, in turns, depends on the type of geological formation. Geoelectric tomography (sometimes indicated as DC tomography, where "DC" means "direct current") is an iterative process aimed at producing a final resistivity model that honors (fits) all the measured electrical potentials along the acquisition layout. The inversion process starts by simulating the electrical response (predicted response) starting from a reasonable initial guess. This is a starting model of the spatial distribution of resistivity based on available prior information. This simulation is known as *forward problem or direct modeling*: a "synthetic" response is derived from an assigned distribution of model parameters. After that step, a *misfit function* is calculated by subtracting the "synthetic" predicted response from the "real" observed response. In the first steps of the inversion process, the misfit is commonly large because the initial guess is just a preliminary resistivity model of the subsoil. Thus, the current resistivity model is updated iteratively to minimize the misfit progressively. The loop continues until the misfit is lowered below a given threshold; this can be, for instance, the standard deviation in the data. A different stopping criterion is ending the loop when further model updates do not produce any significant reduction in the misfit (stationary condition). Model updates are performed using well-known "optimization algorithms," such as conjugate gradient, Newton, or other methods. The final model is a spatial resistivity distribution that allows reproducing, or fitting (by forward simulation) the experimental observations (real data). That model represents *a solution* of the *inverse problem* (i.e., estimating the model parameters from the observed response).

The same inversion approach is applied also in other types of geophysical imaging techniques. For instance, travel time tomography is a well-known approach for retrieving models of seismic velocities from the travel times of seismic waves measured by an array of geophones. Other imaging techniques are based on inversion of seismic impedance, inversion of electromagnetic fields, inversion of gravity data, and so forth (Dell'Aversana, 2014a).

Inversion procedures are applied in medical imaging too, such as in radiography (Stanley, 1983). The problem to solve is to reconstruct an image of a "target" inside the human body, using minimally invasive measurements. Like in geophysics, the measurements must be related to the quantities of interest by known mathematical relationships. To create the image, an X-ray generator produces a heterogeneous beam of X-rays. A certain amount of X-ray is absorbed by the "target" (scanned object). The amount of energy absorbed depends on the particular density and composition of that object. X-ray computed tomography (X-ray CT) makes use of computer-processed combinations of many X-ray images taken from different angles to produce cross-sectional (tomographic) images of a scanned object.[14] As in geophysics, the final radiography image is the result of an inversion process.[15] Of course, X-ray CT requires

---

[14] X-ray CT offers a good example of how technology emulates in some way our basic cognitive processes. The combination of many X-ray beams scanning an object from different angles and the final production of an image from many projections is analogous to the natural process of visual perception. Indeed, when we look to the same object from many points of view, many images of it are formed on our visual cortex. The retina serves as a transducer for the conversion of patterns of light into neuronal signals. These signals are processed in a hierarchical fashion by different parts of the brain. The totality of the visual scene is finally accomplished by the visual association cortex. Our brain performs an extraordinary work of image reconstruction combining a huge quantity of visual information with other perceptive information (such as sounds) and with previous knowledge saved in our memory. Our actions and movements provide a continuous feedback to our multisensory perception.

[15] The mathematical theory behind computed tomographic reconstruction is based on Radon transform by which a function can be reconstructed from an infinite set of its projections. Another reconstruction method is the "algebraic reconstruction technique (ART)" based on approximate solution to a large system of linear algebraic equations.

a mathematical equation relating X-ray absorption to the unknown physical parameters (such as density properties inside the object). This is an exponential law: $I(t) = I_0 \cdot e^{-\mu t}$, where I_0 is the initial intensity of the X-ray beam, $I(t)$ is the intensity of the X-ray beam after traveling through the object for a distance t, and μ is the attenuation coefficient depending on tissue density and on the energy of the beam.

An analogous computation approach based on mathematical inversion is applied to other methods (Hsieh, 2009; Kac and Slaney, 1988) for imaging human tissues, such as MRI, PET, and SPECT. All these techniques work on the same general imaging principle: the human body is scanned by means of some type of radiation transmitted, reflected, or emitted by the body. This produces an "observed response" from which an image of some physical properties of the body is retrieved by inversion.

4.4.3 Similar Problems

X-ray tomography represents a good starting point for introducing some crucial questions related to image reconstruction by mathematical inversion of large amount of observations. This problem is very general and affects different scientific fields, including geophysics.

3D X-ray imaging allows acquiring many projection images of a "target" from different directions. The work of Radon (1917) demonstrates that the inner structure of the target can be determined if these projection images are available from all around a 2D slice of the target itself. Unfortunately, the majority of inverse problem to solve real applications, including biomedical imaging, are typically *ill-posed*. This mathematical term means that the solution is strongly sensitive to small variations in the data.[16] Experimental measurements are commonly affected by noise. This can create large oscillations in the solution of the associated inverse problem. Consequently, when we try to invert the observations for reconstructing the image of a target, we have to solve an ill-posed problem. The practical consequence of inverting noisy data is that the final solution can contain artifacts and can be unreliable. However, ill-posed problems can still be solved satisfactorily if our mathematical formulation includes a model of the noise affecting the data and, possibly, a prior knowledge about the possible solution.

An additional problem in some medical applications is that the ideal full X-ray coverage is not possible, such as in mammography. In that case,

---

[16] The opposite of an *ill-posed inverse problem* is indicated as a *well-posed inverse problem*. That mathematical term was defined by Jacques Hadamard (1902). He postulated the following three conditions for having a well-posed problem: (1) a solution exists; (2) the solution is unique; and (3) the solution's behavior changes continuously with the initial conditions.

the projections can be obtained only from a relatively narrow aperture. Therefore, the ray sampling is not uniform and is affected by shadow zones. This problem is known as *limited-angle tomography*, and the data set is called *sparse projection data*. Unfortunately, traditional CT reconstruction algorithms can produce only low-quality reconstructions when applied to sparse projection data.

Another problem is that the solution should be produced in almost real time. This is a fundamental requirement in medical imaging. The problem is that the mathematical formulation describing body-radiation interaction is based on complex nonlinear equations. The solution of a nonlinear inverse problem applied to huge amount of data commonly requires long computation times (hours or days) and powerful computer resources. Consequently, to make these problems affordable, it is necessary to develop good linear approximations of the nonlinear systems of equations. The final objective is to obtain reliable images in almost real time, which is a very challenging goal.

In summary, medical imaging commonly requires to solve nonlinear, ill-posed inverse problems, in real time, applied to huge data set formed by sparse projection observations affected by some type of noise. These types of problems affect almost all the imaging algorithms applied to real observations measured in different scientific domains, including geophysics. For instance, seismic tomography is a typical nonlinear inverse problem. Analogously to medical imaging, in geophysical imaging, large systems of nonlinear equations must be solved for inverting huge 3D data sets in reasonable times (Zhdanov, 2010; Zhdanov and Fang, 1996). Furthermore, when both source and receiver arrays are deployed at surface, seismic or electromagnetic coverage (i.e., the sampling of the subsoil by seismic or electromagnetic waves) can be very irregular, depending on geological complexity. The final data set can be very sparse and affected by significant noise, thus the problem is strongly ill-posed and the final tomography image can be unreliable. The same happens in gravity and magnetic data inversion in geophysical prospecting. What strategies are used in both medical and geophysical applications for solving all these difficult problems?

4.4.4 Similar Solutions

Statistical inversion is one of the best approaches for both geophysical imaging (Tarantola, 2005) and medical imaging (Siltanen et al., 2003). It allows obtaining robust tomographic reconstructions from sparse collection of projection data affected by noise, including cone-beam geometry and truncated projections. Moreover, statistical inversion allows introducing a priori information about the target. This approach allows transforming the classically ill-posed problem into a well-posed stochastic form (see Appendix 1).

The basic idea in statistical inversion is to formulate the inverse problem in terms of *Bayesian inference*. All variables are redefined to be random variables, characterized by a certain probability distribution. Moreover, the distribution of uncertainties on both observations and model parameters is included in the statistic formulation of the inverse problem. The well-known Bayes formula (that will be discussed in Chapter 6) allows calculating the conditional probability distribution (also called posterior density) of the model parameters, given a set of experimental measurements. That posterior distribution represents the complete solution of the inverse problem. In fact, it expresses our belief of the distribution of model parameters based on all prior information and the measurements, including their uncertainties. Because the posterior distribution is a probability density in a large-dimensional model space, we must have efficient tools to explore it. In other words, we must search for the "best" distribution of model parameters fitting the data. The most common approaches are to search for the maximum a posteriori estimate and conditional mean estimate.

The search methods used in X-ray tomography are the same as those used in the majority of geophysical inverse problems. The most crucial tasks in statistical inversion are the determination of a reasonable prior model, its quantitative "codification" into the language of probability densities, quantification and inclusion of noise in the inversion process, linearization of the equations linking data and model spaces, and the choice of the most appropriate optimization algorithm for minimizing the differences between observations and predicted responses. The Bayesian approach allows formalizing all these aspects. The problem of introducing a prior knowledge can be solved with different strategies. For instance, if the image is expected to be smooth, it is possible to introduce a smoothness condition in the process of minimization itself. This method offers the benefit of stabilizing the whole inversion process and avoiding sharp variations in the spatial distribution of the model parameters. On the other side, if we expect to find local anomalies and/or sharp boundaries between different portions of the final images, the smoothing prior assumption is not appropriate. In fact, in both medical and geophysical imaging, observing sharp contrasts in physical or biological properties is the most common scenario. The Bayes approach is well suited for dealing also with this type of situation through the introduction of *structural priors*. That approach is based on the fact that, in medical imaging as well as in geophysical imaging, the main geometrical features of the target are often well understood.

A prior knowledge can be obtained from different types of measurements. In medicine, it can be derived from anatomical information or from other imaging modalities. In geosciences, structural information about the target can derive from different geophysical methods. For instance, gravity or electromagnetic data are inverted starting from a prior geometrical knowledge derived from seismic data. The key point is that in both medicine and geosciences, information related to different physical parameters is often correlated. For instance, mass absorption coefficient, conductivity, and thermal parameters can show strong correlations because they are related to the same organs in the body. For different geological reasons, parameters such as seismic velocity, electrical conductivity, and density can be very well correlated because they are related to the same geological formations.

Consequently, we can expect that there are *jumps* with respect to all parameters through the organ boundaries or through rock formation boundaries. Hence, this information can be used as structural prior information in the Bayes formulation of an imaging problem in both medical and geophysical applications. The mathematical formalizations of that a prior assumption can be obtained through a proper definition of *model covariance matrix*. We can set this matrix imposing that pixel values of a physical parameter within the same type of tissue (or the same type of geological layer) are strongly correlated. On the other side, the correlation across the boundaries is kept low or negligible.

Another effective approach for solving ill-posed inverse problems is *simultaneous joint inversion* of the measurements obtained by independent methods but linked in the model space (see Appendix 2). For instance, in geophysics, seismic and electromagnetic measurements can be linked through the physical parameters of porosity and fluid saturations. In fact, both seismic and electromagnetic responses depend on these rock properties. Empirical relationships linking these physical parameters are well known in the literature of rock physics. This approach represents a powerful method for combining different imaging technologies. Bayesian joint inversion allows producing high-resolution images in both geophysical and medical fields.[17]

Indeed, increasing efforts are dedicated in geosciences, and in particular in the hydrocarbon industry, for developing hybrid acquisition methodologies and algorithms of Bayesian joint inversion of multidisciplinary geophysical data. Many practical results have been obtained over the past few years (Dell'Aversana, 2014a).

On the other side, an interesting trend in various medical fields is the application of a similar integrated approach. For instance, hybrid methods combine ultrasound and electromagnetic data. The process of deriving useful information from multimodality medical images is called "medical image fusion." The integrated information can be used for many

---

[17] I will discuss some details of this technology in Chapter 7 and in Appendix 2.

different purposes including diagnosing diseases and surgery intervention. As it happens in geophysics, the final integrated images add significant diagnostic value with respect to any single modality image. Of course, combining different images obtained by independent methodologies such as CT, MRI, and PET into a single image is not an easy task.

In summary, geophysical and medical imaging represent fields of study and applications that share similar problems, approaches, and algorithms. Because of the intrinsic difficulties of the imaging problems in both domains, it is reasonable to encourage the cooperation between experts for creating multidisciplinary teams formed by researchers of both scientific fields.

4.4.5 Expanding the Concept of Imaging

In this chapter, I have explained that mental maps are formed in response to every type of sensory stimuli, and not exclusively to visual perception. It is true that in human brain there is a sort of "visual dominance" with respect to the other senses. However, multiple researches confirm that the auditory perception of spatial location can predominate on vision when visual stimuli are degraded or ambiguous (Alais and Burr, 2004). Multisensory integration is the rule and not the exception in basic perception and in high-level cognition. That is true especially when each constituent individual sensory stimulus arises from approximately the same location. Damasio (2010) explains very well that sensory integration is basically a question of combinations of correlated neural maps. That integration process arises in both cortical layers and in specific subcortical nuclei. For instance, audiovisual cross-modal interactions occur in the auditory association cortex near the Sylvian fissure[18] in the temporal lobe. Moreover, the dorsal auditory pathway projecting from the temporal lobe is related to spatial information processing.

I have already remarked that superior colliculus plays a key role in integrating multisensory information. In fact, receptive fields from somatosensory, visual, and auditory perceptions converge in its deeper layers creating 2D multisensory maps of the external world.

In summary, imagery derives from a complex process of correlation of different types of neural maps in different parts of the brain. It represents an *integrated neurobiological process* related to multisensory perception.[19]

---

[18] The Sylvian fissure, also known as the lateral sulcus, separates the frontal and parietal lobes superiorly from the temporal lobe inferiorly.

[19] I remark that perception of the external world is constantly combined with perception of our internal state, with previous knowledge saved in our memory, with our emotions, expectations, and so forth. Thus, neural maps are influenced also by internal variables concerning the interior of our body.

Multisensory or multimodal perception has been widely studied in cognitive sciences and neurosciences. After the pioneering studies of Stratton (1896) many steps forward have been done. Nowadays, a huge number of experimental evidences confirm that the perception of a stimulus is influenced by the presence of another type of stimulus. For instance, visual acuity can be improved by auditory stimuli. Furthermore, the cognitive process of "selective attention" can be improved significantly when multiple senses are focused on the same object. In general, responses to multiple simultaneous sensory stimuli can be faster and more effective than responses to the same stimuli presented in isolation. In summary, integrated maps and multisensory neurons are continuously activated in the brain for connecting multisensory stimuli into a single percept.

Based on that neurobiological background, we can generalize the concept of imaging. *Expanded imaging* is an integrated and coherent representation of the world based on multisensory mental maps. This concept of imaging enlarged to other senses has important practical implications. For instance, sonification[20] and multimodal display of physical signals (simultaneous representation of images and sounds) have found significant applications in many fields of study, including geophysics and medicine. Sonification has been used in geosciences (Benioff, 1953; Speeth, 1961; Hayward, 1994; Saue, 2000; Kilb et al., 2012; Peng et al., 2012; Quintero, 2013) and in other fields (see Dubus and Bresin, 2013 for a recent and exhaustive review) for several decades. Recently, Dell'Aversana (2014b) demonstrated how multimodal display of seismic data can support exploration geosciences by detection and classification of important seismic signals, including gas-filled channels, faults, and stratigraphic features.

Sonification has many applications in medical diagnosis (Fitch and Kramer, 1994), for analysis of heart sound (Pickover, 1991) and audio display of other vital information such as respiration and blood pressure. Furthermore, multimodal display is used as a valid support for positioning and guidance of instruments during surgery intervention, image interpretation during EEG, supporting the analysis of PET scans of the human brain, and diagnosis of moderate to severe Alzheimer's disease.

---

[20] Auditory perception and multimodal display usually involve the concept of sonification. This is a set of techniques used in several research fields to transform data into sounds and to represent, convey, and interpret them. Auditory perception can be a useful complement to visualization techniques. In fact, it offers many perceptive and cognitive benefits in terms of temporal, spatial, amplitude, and frequency resolution.

In the third part of this book dedicated to brain-based technology, I will discuss methodological and applicative aspects of multimodal analysis (images and sounds) of geophysical data.

4.5 SUMMARY AND FINAL REMARKS

Exploration is strongly based on mapping and imaging. When geophysicists think about imaging, they commonly intend a set of algorithms and techniques aimed at creating pseudoimages of the subsoil, such as seismic sections or 3D volumes of some geophysical attribute. With the support of those images, they can take practical decisions in the exploration workflow, such as optimal positioning of a well in hydrocarbon research.

In this chapter, I showed that the question of imaging is not confined to mere technological aspects. It has deep historical, epistemological, and neurobiological roots and concerns cognitive and exploratory attitudes of humans. The fact that imaging is so important in exploration depends on the structure of our brain: its architecture is organized for creating and manipulating images, even in absence of external stimuli. Mental imagery, sometimes confused with "imagination," is the ability to form images in the mind. Imagery is a cognitive feature of almost all human beings (excluding those with specific brain dysfunctions). Instead, I consider *imagination* as a particularly creative mental imagery attitude, typical of people with great fantasy.

Imagery is so important because mental images are the fundamental *bricks* for representing the internal and the external world and, finally, for thinking and reasoning. Mental images do not represent any esoteric concept. Indeed, they have a neurobiological background. They correspond to complex patterns of neural connections formed on the neocortex and on some specific subcortical nuclei. Imagery is an activity deeply rooted in our brain.

Mapping is strictly related to imaging, imagery, and imagination. The process of mapping allows visual representation of complex information, producing a synoptic view of huge amount of data. It is a way to create images of "what we know" and, sometimes, of "what we believe to know."

The same cognitive and theoretical background of imaging and mapping is shared by many scientific domains. Consequently, it is reasonable to expect analogous problems and similar solutions in different scientific fields. In this chapter, I discussed briefly Bayesian inversion (additional details are provided in Appendices 1 and 2). This represents a smart approach for creating robust models and images in geophysics and in diagnostic medicine. It offers the great benefit to incorporate in the same inversion process our actual knowledge (measurements), our assumed knowledge (a prior model), and the uncertainties on both types of information.

Despite their importance, images represent just one aspect of imagery. In fact, this cognitive process is related to multisensory perception. Not only visual stimuli but also other sensory modalities concur to create brain maps and mental images. The concept of imaging can be generalized taking into account the ability of our brain to combine different types of mental maps. *Expanded imaging* is an integrated and coherent representation of the world based on multisensory mental maps. This expanded view of imaging and imagery leads directly to the development of hybrid technologies based on simultaneous audio and video display. Finally, this multisensory approach allows improving analysis of complex information.

I would like to conclude this chapter with some further remarks about the key role of imagination in exploration geosciences. Indeed, I think imagination is fundamental for every successful exploration project. This is a general rule, not confined to geosciences.

The European explorers Cristoforo Colombo, Vasco da Gama, and, later, James Cook, discovered new continents, thanks to their uncommon imagination attitude combined to their exceptional ability as cartographers. The famous captain Cook (1728—79) of the English Royal Navy made detailed maps of Newfoundland and much of the entrance to the Saint Lawrence River during the siege of Quebec, before sailing thousands of miles across largely uncharted areas of the world. He traveled through the Pacific Ocean, mapping on a scale not previously achieved, the eastern coastline of Australia and New Zealand. James Cook was a perfect example of combination of skills typical of a seaman, an explorer, and a cartographer.

Exploring the Earth interior requires the same basic attitudes as exploring the globe surface. It requires the ability to create detailed maps of geological outcrops, high-resolution images of the subsoil, and, finally, to link surficial and deep features. Mapping and imaging technology is fundamental for that activity. However, I think that creating a good geological model would be impossible without mental imagery and imagination. This ability is particularly pronounced in geologists and geophysicists. Indeed, it is not without good reasons that Charles Lyell introduced a *Theory of the geological imagination* in the first volume of *Principles of Geology.* He understood that imagination is necessary for transforming geological images and maps of surficial observations into dynamic models of the Earth interior.

References

Alais, D., Burr, D., 2004. The ventriloquist effect results from near-optimal bimodal integration. Curr. Biol. 14 (3), 257–262. http://dx.doi.org/10.1016/j.cub.2004.01.029. PMID 14761661.

Benioff, H., 1953. Earthquakes around the world. In: Cook, E. (Ed.), On Out of This World. Side 2.

Damasio, A., 2010. Self Comes to Mind: Constructing the Conscious Brain. Pantheon, New York.

Dell'Aversana, P., 2013. Cognition in Geosciences: The Feeding Loop Between Geo-Disciplines, Cognitive Sciences and Epistemology. EAGE Publications. Elsevier.

Dell'Aversana, P., 2014a. Integrated Geophysical Models: Combining Rock Physics with Seismic, Electromagnetic and Gravity Data. EAGE Publications.

Dell'Aversana, P., 2014b. A bridge between geophysics and digital music. Applications to hydrocarbon exploration. First Break 32 (5), 51–56.

Dell'Aversana, P., July 2015. The brain behind the scenes — neurobiological background of exploration geophysics. First Break 33, 41–47.

Dennett, D., 1969. The Nature of Images and the Introspective Trap, in Block 1981.

Dennett, D., 1979. Two Approaches to Mental Images, in Block 1981.

Dubus, G., Bresin, R., 2013. A systematic review of mapping strategies for the sonification of physical quantities. PLoS One 8 (12), e82491. http://dx.doi.org/10.1371/journal.pone.0082491.

Fitch, T., Kramer, G., 1994. Sonifying the body electric: superiority of an auditory over a visual display in a complex, multivariate system. In: Kramer, G. (Ed.), Auditory Display: Sonification, Audification and Auditory Interfaces. Addison-Wesley, Reading, MA.

Freeman, W.J., 1999. How Brains Make up Their Minds. Weidenfeld & Nicolson.

Freeman, W.J., 2000. Neurodynamics. An Exploration in Mesoscopic Brain Dynamics. Springer, London.

Freeman, W.J., March 2007. Definitions of state variables and state space for brain-computer interface. Cogn. Neurodyn. 1 (1), 3–14. http://dx.doi.org/10.1007/s11571-006-9001-x. Published online 2006 December 7.

Freeman, W.J., 1987. Simulation of chaotic EEG patterns with a dynamic model of the olfactory system. Biol. Cybern. 56, 139–150.

Freeman, W.J., Vitiello, G., 2006. Nonlinear brain dynamics as macroscopic manifestation of underlying many-body field dynamics. Phys. Life Rev. 3, 93–118.

Freeman, W.J., Rogers, L.J., Holmes, M.D., Silbergeld, D.L., 2000. Spatial spectral analysis of human electrocorticograms including the alpha and gamma bands. J. Neurosci. Meth. 95, 111–121.

Hadamard, J., 1902. Sur les problèmes aux dérivées partielles et leur signification physique. Princeton University Bulletin, pp. 49–52.

Hayward, C., 1994. Listening to the Earth sing. In: Kramer, G. (Ed.), Auditory Display: Sonification, Audification, and Auditory Interfaces. Addison-Wesley, Reading, MA, pp. 369–404.

Hsieh, J., 2009. Computed Tomography Principles, Design, Artefacts and Recent Advances, second ed. SPIE, Bellingham.

Huth, G.H., de Heer, W.A., Griffiths, T.L., Theunissen, F.E., Gallant, J.L., 2016. Natural speech reveals the semantic maps that tile human cerebral cortex. Nature 532, 453–458. http://dx.doi.org/10.1038/nature17637.

Kac, A., Slaney, M., 1988. Principles of Computerized Tomographic Imaging. IEEE Press.

Kilb, D., Peng, Z., Simpson, D., Michael, A., Fisher, M., Rohrlick, D., March/April 2012. Listen, Watch, Learn: SeisSound Video Products, Electronic Seismologist.

Menke, W., 1989. Geophysical Data Analysis: Discrete Inverse Theory. Academic Press.

Nabighian, M.N., 2008. Electromagnetic Methods in Applied Geophysics: Applications, Part A and Part B. Society of Exploration Geophysicists.

Peng, Z., Aiken, C., Kilb, D., Shelly, D.R., Bogdan, E., 2012. Listening to the 2011 Magnitude 9.0 Tohoku-oki, Japan, Earthquake, Electronic Seismologist, N. of March/April. Cook Laboratories, Stamford, CT, 5012 (LP record audio recording).

Pickover, C.A., 1991. A note on the visualization of heart sounds. Leonardo 24 (3), 359–361.

Pylyshyn, Z., 1973. What the mind's eye tells the mind's brain—a critique of mental imagery. Psychol. Bull. 80, 1–24.

Pylyshyn, Z., 1978. Imagery and artificial intelligence. In: Wade Savage, C. (Ed.), Perception and Cognition: Issues in the Foundation of Psychology. University of Minnesota Press, Minneapolis, pp. 19–56.

Quintero, G., 2013. Sonificaton of oil and gas wire line well logs. In: International Conference on Auditory Display (ICAD 2013).

Radon, J., 1917. Über die Bestimmung von Funktionen durch ihre Integralwerte längs gewisser Mannigfaltigkeiten. In: Berichte über die Verhandlungen der Königlich-Sächsischen Akademie der Wissenschaften zu Leipzig, Mathematisch-Physische Klasse [Reports on the Proceedings of the Royal Saxonian Academy of Sciences at Leipzig, Mathematical and Physical Section]. Leipzig, Teubner (69), 262–277; Translation: Radon, J.; Parks, P.C. (trans.) (1986), On the determination of functions from their integral values along certain manifolds, IEEE Trans. Med. Imaging 5(4), 170–176. doi:10.1109/TMI.1986.4307775. PMID 18244009.

Saue, S., 2000. A model for interaction in exploratory sonification displays. In: Proceedings of the 6th International Conference on Auditory Display (ICAD 2000). Atlanta, GA, USA, pp. 105–110.

Siltanen, S., Kolehmainen, V., Järvenpää, S., Kaipio, J.P., Koistinen, P., Lassas, M., Pirttilä, J., Somersalo, E., May 21, 2003. Statistical inversion for X-ray tomography with few radiographs I: general theory. Phys. Med. Biol. 48 (10), 1437–1463.

Speeth, S.D., 1961. Seismometer sounds. J. Acoust. Soc. Am. 33, 909–916.

Stanley, R.D., 1983. The Radon Transform and Some of its Applications. John Wiley and Sons, New York (Republished by Dover Edition in 2007, as a Revised Edition of the Previous Krieger Publishing Company in 1993).

Stratton, G.M., 1896. Some preliminary experiments on vision without inversion of the retinal image. Psychol. Rev. 3 (6), 611–617. http://dx.doi.org/10.1037/h0072918.

Tarantola, A., 2005. Inverse problem theory and methods for model parameter estimation. Soc. Ind. Appl. Math. ISBN: 978-0-89871-572-9.

Yilmaz, Ö., 2001. Seismic Data Analysis: Processing, Inversion, and Interpretation of Seismic Data. SEG Books.

Zhdanov, M.S., 2009. Geophysical Electromagnetic Theory and Methods. In: Methods in Geochemistry and Geophysics, vol. 43. Elsevier.

Zhdanov, M.S., 2010. Electromagnetic geophysics: notes from the past and the road ahead. Geophysics 75 (5), 75A49–75A66.

Zhdanov, M.S., Fang, S., 1996. Quasi-linear approximation in 3-D E. M. modeling. Geophysics 61, 646–665.

Web References

Image Library of National Institute of Mental Health, image in the public domain: https://images.nimh.nih.gov/public_il/searchresults.cfm.

5

Recognition

5.1 GEOPHYSICAL PROBLEMS

Since 1915, Alfred Wegener observed how the east coast of South America and the west coast of Africa looked as if they were once attached. Wegener supported his idea that the continents had once been joined, and over time had drifted apart, with many paleontological and climatological evidences. In 1950, an extraordinary example of pattern recognition confirmed the theory. Indeed, magnetic variations observed on the seafloor revealed recognizable patterns. Alternating stripes of magnetically different rocks appeared distributed in rows on either side of the mid-ocean ridge, supporting the idea of opening oceans and moving continental plates. Over the following years, additional observations of stratigraphy, paleontology, and structural geology confirmed the hypothesis that many continental margins, now separated by thousands of kilometers, were joined originally.

The plate tectonic theory represents the most famous example of how geosciences are frequently based on recognition of patterns and "meaningful shapes" at extremely variable scale. Almost all the sectors of modern Earth disciplines are based on recognition of patterns. For instance, geomorphology is the scientific study of the landscape. It investigates the origin and evolution of topographic and bathymetric features. These are natural patterns and shapes determined by physical, chemical, and biological processes. In sedimentology, the study of the shape of particles is fundamental for inferring physical and chemical processes such as erosion, weathering, transport, deposition, and diagenesis. Analysis of shapes is crucial also at small scale (often with large-scale implications), in mineralogy, petrography, and paleontology, for reconstructing geodynamic processes, environmental energetic conditions, and paleogeography of entire regions. The study of the shape of volcanoes is crucial for reconstructing the eruptive history in a volcanic region.

Pattern recognition approaches play a key role in exploration geophysics. In hydrocarbon research, sequence stratigraphy and interpretation of structural patterns provide insight into basin's petroleum system. Seismic facies analysis is the study of configurations of seismic properties such as reflection, amplitude, and frequency. It is very useful to determine changes in lithology, fluid types, sedimentary, and structural features at variable scale. The basic principle underlying seismic sequence analysis is that the seismic volume can be partitioned into "packages" of relatively conformable reflections. These are bound by unconformities clearly identifiable by reflection terminations and/or discordant surfaces. The patterns of reflections are interpreted in terms of geological units with the help of well data and composite logs.

Another important approach used in reflection seismology and based on pattern analysis is amplitude versus offset (AVO) or amplitude variation with offset. It is also known as amplitude versus angle. This technique uses the dependency of the seismic amplitude with the distance between source and receiver (the offset). AVO helps to determine rock's porosity, density, or seismic velocity and information about fluids (such as hydrocarbons). An AVO anomaly is commonly expressed as patterns of increasing or decreasing reflected amplitudes with offset, depending on the contrast between elastic properties at the reflecting interface. This is a typical example of how geophysicists use patterns of signals for inferring information about rock properties and geology.

Another well-known example of geophysical pattern related to fluids in the rocks is the "flat spot." This seismic attribute anomaly appears as a horizontal reflector cutting across the stratigraphy on the seismic image. It can indicate the presence of hydrocarbons. In fact, it can be caused by the increase in acoustic impedance when a gas-filled porous rock overlies a liquid-filled porous rock. However, it could be determined by other causes, such as a mineralogical change in the rocks.

Pattern recognition is important also in other geophysical approaches that are different from reflection seismic. For instance, in electromagnetic prospecting (marine controlled source electromagnetics), a technique similar to seismic AVO is used as an indicator of resistivity anomalies. These can be interpreted as variations in lithology or in fluids. Consequently, this approach finds many applications in hydrocarbon exploration.

The shape of gravity and magnetic data measured at surface is equally important for inferring information about the depth and the nature of density and magnetic causal bodies.

At a smaller scale, patterns of geo-radar (or ground penetrating radar) reflections can indicate the presence of buried objects. Experienced geophysicists are able to distinguish different types of targets just looking at the reflection patterns appearing on the computer display in real time (as discussed in the case history of Chapter 1).

In a completely different field, log analysts compare and cross-plot multiple measurements made in a well (clay content, electric conductivity, sonic velocity, and so forth). Different patterns on the "scatter plot display" can be clustered and allowed to characterize different geological formations.

Many other examples of geophysical/geological pattern recognition can be extracted from the practice in exploration geosciences. The key concept is always the same: geoscientists try to *arrange* their observations to facilitate the search of some type of "shape," or pattern, with a geological meaning.[1] Indeed, like imaging and mapping, pattern recognition represents a key attitude deeply rooted in the mammalian brain.

5.2 NEUROBIOLOGICAL BACKGROUND

Our brain never perceives an image as stand-alone information. Every day we combine millions of images and interpret them in the frame of a context. This can include infinity of other types of information forming a complex and dynamic background. Moreover, our previous experiences saved in memory concur to our interpretation. Every image is like a word in a sentence: each individual term has significance only in the context of

---

[1] Seismic processing, aimed at creating stacked/migrated data volumes, is a good example of how field data are combined into structured images. Interpreter geophysicists analyze these images for finding patterns and features with a geological meaning, such as faults, sedimentary contacts, and so forth.

the whole sentence. On the other side, the sentence itself makes sense only if it is formed by meaningful words satisfying syntactical and grammar rules. The same type of "semantic loop" arises for images. They gain significance if they form a recognizable meaningful pattern in the frame of a wide scenario.

In the same way as it happens in the ordinary life, images in geosciences are interpreted as patterns in the context of a geological scenario. The eyes of the interpreter geophysicists continuously search for images that form shapes and patterns with a geological meaning. This situation is well known to geoscientists who interpret seismic sections. Detailed analysis performed on few records can be useful. However, that analysis never stops to a single seismic trace. Indeed, geoscientists look for patterns of reflections forming seismic horizons in ensembles of traces. These "seismic events" represent patterns of reflections (or of diffractions) interpretable in the frame of a geological background. In general, all geoscientists look for patterns. These can appear at surface from the correlation between geological outcrops or from the comparative analysis of associations of fossils. Finally, these patterns are interpreted for reconstructing the paleogeography of a region, creating a geological model of a hydrocarbon reservoir, and so on. The same happens in medical sciences, where images are interpreted in terms of patterns included in a wide medical scenario. The final diagnosis is commonly based on associations of signs, blood parameters, images of cells, tissues, and organs.

A general aspect of patterns is their intrinsic hierarchical nature. For instance, images can be clustered and organized into larger images, forming patterns with increasing complexity. Furthermore, these individual patterns can form *clusters of patterns*, creating larger and larger "structured images." Again, the analogy with language can illuminate the hierarchical structure of patterns: single letters form words, words form sentences, sentences form paragraphs, and so forth. The hierarchical structure of our language and our thought reflects a similar structure in our brain and nature (Dell'Aversana, 2013a). Kitzbichler et al. (2009) from the University of Cambridge discovered that our cognition has intrinsic fractal features.[2] In fact, the amount of time that different brain regions

---

[2] Fractal objects show typically self-similar patterns that can be properly described using a power low relationship. This means they are self-similar at different scales because of a pattern repeating itself. Many natural systems tend to assume a fractal organization on their own, with the advantage that communication and energy distribution are both facilitated. A fractal structure allows a good internal interaction between individual parts and, at the same time, between these parts and the external environment. In general, branch-like patterns easily connect to other systems. Finally, the fractal structure does not depend critically on any single component. This characteristic makes the whole organism/organization stable and not subjected to local perturbations.

spend in synchronized patterns has a power law distribution: among the neural connections, there are many short linkage times and fewer long ones. Moreover, the spread of neural connections across different brain regions has a branch-like distribution. Also the "external nature," including geological phenomena, shows fractal geometry (Mandelbrot, 1967, 1982). Turcotte (1997) shows, with many examples, how fractal geometry can properly describe many processes/systems such as drainage networks and erosion, mineral and petroleum resources, mantle convection, and magnetic field generation. A satellite view is sufficient for observing many examples of fractals, such as river networks and morphology features of mountain ranges.

There is a deep reason why pattern recognition is a distinctive feature of biological systems: recognition represents an effective way to learn from experience. That is true also at a basic biological level, starting from a molecular scale. Gerald M. Edelman suggests that in biology there are many types of "natural recognition systems." Two of these are the immune system and the brain (Edelman, 1987, 1992). The immune system is a selective system of vertebrates consisting of specialized molecules, cells, and organs. It is able to recognize the difference between "self" and "not-self" at a molecular scale. The organism develops an immune system since the birth. That system improves during the entire life in response to various external biological attacks.

In his theory of "selective neural group," Edelman assumes that the brain works also as a selective recognition system. In fact, a "primary repertory" (innate set) of neural connections exists since birth. However, only those connections that allow an adaptive behavior (improving survival probability) are reinforced over time. Moreover, new neural groups and additional connections between different groups are continuously created over a lifetime, in response to the same adaptive selective criteria. Edelman called "neural Darwinism" that process. It is a process of continuous recognition of environmental variability, including visual as well as acoustic patterns, and any other significant sensory stimulus. The brain changes continuously to select and reinforce the neural connections correspondent with an adaptive response. That response consists of two main aspects: adequate internal regulation, called homeostasis (regulation in temperature, pressure, etc.), and adequate external behavior. In summary, both the immune system and the brain of the vertebrates, including human beings, evolved for operating a continuous process of recognition and selection, with final adaptive objectives.

Mark P. Mattson remarks very well the role of pattern recognition in human cognition. He says: "The fundamental function of the brains of all animals is to encode and integrate information acquired from the environment through sensory inputs, and then generate adaptive behavioral responses. Sensory information is first rapidly encoded as patterns inherent in the inputs, with visual and auditory patterns being most extensively studied in mammals ... The large numbers of encoded images and sound patterns can then be recalled and mentally manipulated in ways that enable comparisons of different patterns and, at least in the human brain, the generation of new patterns that convey objects and processes that could possibly exist, or are impossible or implausible" (Mattson, 2014).

These words summarize very well the concept that *cognition is based on recognition*. This is strictly linked with the other cognitive functions that I have discussed in the previous chapter: mapping and imaging. Cognitive maps of the physical environment require a process of encoding and recalling of locations of resources (food, refuge, etc.), potential predators, and navigation landmarks.

The hippocampus plays a key role in this instinctive "recalling and mapping attitude" (Pearce et al., 1998; Spiers et al., 2001). In fact, the visual and auditory patterns (as well as the other sensory inputs) are transferred into hippocampal circuits via neurons in the entorhinal cortex.[3] This is an area of the brain located in the medial temporal lobe and represents the main gateway between the hippocampal formation and the neocortex. It works as a sort of hub in a widespread network for memory and navigation. Furthermore, it has been observed that neurons in the dentate gyrus of the hippocampus play a particularly important role in spatial pattern separation in mammals (Gilbert et al., 1998).

Other regions of the brain involved in recognition are the visual cortex, the prefrontal cortex, and the parietal−occipital−temporal juncture. Processing complex visual and auditory patterns is particularly important in humans because of their centrality to language, imagery, and imagination (Mattson, 2014).

Many other neuroscientists, philosophers, and experts of artificial intelligence have supported the importance of hierarchical pattern processing for explaining the main aspects of thought. Among the others, I mention (in a very synthetic way) the point of view of the inventor and futurist writer Ray Kurzweil (2012). He states that the brain functioning is based on a fractal structure. The neocortex contains 300 million "general pattern recognition circuits" responsible for the main aspects of human cognition. Kurzweil (supported by evidences in neurosciences) says that the whole brain can evolve by repeated exposure to patterns following a basic principle of self-organization.

I must remark that the pattern recognition theory of mind proposed by Kurzweil has been criticized by several philosophers and cognitive scientists for many different reasons. Indeed, human brain does more than

---

[3] The name entorhinal (inside rhinal) cortex derives from the fact that it is partially enclosed by the rhinal (olfactory) sulcus.

comparing and matching patterns of external stimuli. However, it is difficult to doubt about the central role of pattern recognition in the functioning of human mind. As stated by Mattson, superior pattern processing seems to be *the essence of the evolved human brain*.

5.3 LINKS AND EXAMPLES

Pattern recognition is an instinctive attitude that we apply almost continuously and unconsciously. Indeed, the word "recognize" derives from Latin word "recognoscere," which means "recall to mind, to know again." The cognitive process of "recalling in mind" is often immediate and does not require any explicit computation. When we recognize a friendly face confused in a multitude of people, we never calculate quantitative features such as the intraocular distance or the nose length. We just recognize that person. The same happens when we recognize a song. Even if it is played with other instruments and with a different style with respect to the original version, we can recognize it without any computation or quantitative analysis.

On the other side, in some types of research we need to apply automatic recognition approaches based on mathematical algorithms working on powerful machines. For instance, in exploration geophysics, data volumes are often so big that it is difficult, or impossible, to examine visually every individual seismic trace. Consequently, automatic classification algorithms are frequently applied to extract features of geological interest. In these cases, the term pattern recognition indicates a quantitative branch of "machine learning" that focuses on the recognition of patterns and regularities in data. This approach is applied for big data mining, automatic classification, information clustering, and many other purposes. Similar problems of automatic classification arise in other scientific sectors, including medical, social, and financial disciplines. Pattern recognition and big data mining approaches are used, for instance, in many business and commercial sectors for attempting to identify patterns in our spending and habits.

In the next section, I will discuss a synthetic comparative study of algorithms and approaches used in two fields only apparently different: geophysics and music. That comparison is justified and motivated by the many analogies existing between geophysics and music. For instance, the physical background of seismology is very similar to the physics of the sound. Consequently, many algorithms, approaches, and technologies can be exported from one domain to another, including machine-learning and pattern recognition methods. There are also robust cognitive motivations for linking the geophysical and the musical domain. Our brain evolved for integrating multisensorial information (cross-modal cognition). Consequently, combining musical and imaging attributes through simultaneous audio–video display can improve the process of anomaly detection and geophysical data interpretation. In the third part of the book, I will expand these concepts showing that they represent the background of innovative brain-based technologies.

5.3.1 Pattern Recognition in Geophysics and in Music

5.3.1.1 Geophysical Domain

In geosciences, recognition and classification happen through qualitative and quantitative approaches. For instance, field geologists and geophysicists who interpret seismic data commonly start from detecting interesting patterns just by visual recognition; then they continue the analysis using quantitative techniques. Qualitative recognition is strongly based on analogical thinking. This is true in all the interpretative sciences, especially where visual perception plays an important role, such as in Earth and medical disciplines. For example, patterns of reflections observed in a seismic data set are commonly considered interesting if they "recall to mind" real geological "objects" observed at surface (Fig. 5.1). These can be faults, structural highs, unconformities, erosional truncations, channels, changes of facies, and other sedimentary/structural features.

When dealing with big 3D data sets, it can be useful (or necessary) to support qualitative interpretation with automatic approaches of data mining, recognition, detection, and classification (Zhao et al., 2015; Duda

FIGURE 5.1   Comparison between fluvial meandering at Earth surface (right) and a seismic slice of channels (left). Pictured here is an RGB blended attribute volume that allows you to scan through in 3D space to identify features characterized by varying seismic responses. *Courtesy: Dr. Gaynor Paton, extracted from Geoteric website: http://www.geoteric.com/geoteric-CI.*

et al., 2001). The term "clustering" is often used as a synonymous of classification. It indicates the operation of grouping patterns of signals showing similar values of one or more selected attributes. For instance, seismic reflections showing similar geometrical features and comparable frequency content can be clustered in the same seismic facies.

The process of pattern recognition in geophysics commonly starts from focusing on a given stratigraphic formation or suite of formations. This part of the workflow is known as *segmentation*; it takes into account apparent facies variations caused by vertical and lateral resolution changes with depth.

Another key step of the workflow is defining the *features* we desire to extract from the data. This step is known as *feature extraction*. A feature is a specific piece of information that can be used for characterizing the data. In geophysics, these are commonly indicated as *attributes*. Geophysical attributes can be related to amplitude, frequency, geometrical reflector configurations, lithology, geomechanical properties, and so forth.

Classification is the process by which each voxel is assigned to one of a finite number of classes (called clusters). The number of classes is commonly decided a priori by the user. He/she can set a finite number of categories in which the data can be grouped, based on their similar features. These a priori categories form a sort of *taxonomy* aimed at organizing the data in discrete classes. Cluster taxonomy is necessarily affected by arbitrary choices. However, it can be supported by well data and robust geological criteria. A good taxonomy should drive the process of pattern recognition toward meaningful clustering results. Ideally, each cluster should represent a seismic facies corresponding to a geologic facies. In practice, this condition is not necessarily honored.

Pattern recognition and data clustering can make use of *unsupervised or supervised learning algorithms*. In the first case, the interpreter provides no prior information other than the selection of attributes and the number of desired clusters. Alternatively, supervised learning classification requires a phase of *training*. The interpreter can use prototypical and selected examples for training purpose. These represent a *training data set* that helps the pattern recognition system to decide what is typical of a particular class.

The final step is *cross-validation*. A simple approach is the hold-out method. The training data is split into two disjoint parts: (1) a new *training subset*, used just to train the classifier algorithm and (2) a *test subset*, used to estimate the error rate of the trained classifier. At the first step, only the training subset is used for retrieving a "model" able to predict the data. This model is often indicated as *function approximator*. At the second step, the function approximator is asked to predict the values for the data belonging to the testing subset. This represents a data subset that is unknown for the approximator itself. The prediction errors are used to evaluate the model reliability (the effectiveness of the approximator). The cross-validation method is usually preferable to the simple computation of a residual function, where the predicted values are compared with the whole data set. In fact, residual evaluations do not give any indication of how well the learner algorithm will work when it is asked to make new predictions for unknown data.

K-fold cross-validation is a generalization of the hold-out method, where the data set is divided into K subsets and the hold-out method is iterated K times.

The leave-one-out cross-validation method is a further expansion of the K-fold cross-validation method, setting K equal to the total number of data points.

Finally, after proper cross-validation, the interpreter decides if the classification work produced reasonable results. For instance, he/she will evaluate whether a given cluster represents a unique seismic facies or whether it should be further subdivided, perhaps selecting different features, excluding or adding additional attributes. Eventually, the training can be performed on a different subset or different classificatory methods can be used.

A large variety of classification methods are applied in completely different scenarios, such as business failure prediction, risk assessment, big data mining, and so on. A good comparative revision has been performed by Kiang (2003). An accurate review in the geophysical domain has been published by Zhao et al. (2015), where different methods are compared. An interesting application of seismic facies recognition in the Canterbury Basin of New Zealand is discussed in that work. Furthermore, the authors provide a useful discussion about the advantages and limitations of each clustering method. Finally, they illustrate critical areas for future algorithm development and workflow refinement.

Clustering methods can be distinguished according to whether the data are clustered into fuzzy or crisp (hard) subsets. Hard clustering methods are based on the principles of classical set theory: an object either does or does not belong to a class. Instead, in the fuzzy clustering methods, each data point can belong to more than one cluster or partition with different *degrees of membership*. An index between 0 and 1 defines the level of partial membership. This approach is more natural than any rigid form of classification (hard clustering).

Many different types of unsupervised and supervised algorithms exist.

The k-means method is a well-known unsupervised learning algorithm. After setting the number of clusters, the means or centers of each cluster are iteratively defined in a multidimensional feature space. Each data point is assigned to the cluster to whose mean it is closest. Adding a new data point will change the cluster mean. Thus, the mean is recomputed, and the process is repeated.

Self-organizing maps (SOMs) represent another set of unsupervised learning techniques. These produce topologically ordered mapping of the clusters with similar classes lying adjacent to each other. SOMs apply competitive learning instead of error-correction learning,[4] preserving the topological properties of the input space.

A simple supervised method is the k-nearest neighbors algorithm [or (k-NN) for short]. It is used for both classification and regression. In classification problems, an object is classified taking into account the properties of its neighbors (commonly forming the training set). The object is assigned to the class most common among its k-NN.

Effective methods of supervised learning classification are based on artificial neural networks (ANNs).[5]. These have been used for lithofacies recognition, formation evaluation from well-logs data, automated picking of seismic events (reflections and first arrivals), identification of causative source in gravity and magnetic applications, pattern recognition and classification in geomorphology, and target detection and big data mining in hydrocarbon exploration.

Support vector machine (SVM) is another category of supervised learning methods. It is used for solving both classification and regression problems. SVM requires finding an optimal *decision surface* that separates objects having different class memberships. When the data lie in a plane, the clusters can be separated by a line. However, most problems of classification require complex hyperplanes (commonly called *decision boundaries*) in a multidimensional space, which separates cases of different class labels. The basic idea of SVMs is to find a hyperplane in the multidimensional feature space that separates the data into classes with an optimal *margin*. This is defined to be the smallest distance between the separation hyperplane and the training vectors (Bishop, 2006).

Another supervised classification method is the naïve Bayes classifier. It is based on the Bayes rule of conditional probability. It uses all the features contained in the data, analyzing them individually and assuming their mutual independency.

As stated by Zhao et al. (2015) and Kiang (2003), each classification approach has benefits and limitations. Comparisons among different algorithms indicate that each method is more suitable for some applications but not for solving every type of classification problem. Indeed,

---

[4] Competitive learning uses artificial neural networks in which nodes compete for the right to respond to a subset of the input data. Instead, error-correction learning uses optimization criteria such as back propagation with gradient descent.

[5] However, ANNs can be used in both unsupervised and supervised analysis. For instance, the multilayer perceptron is a type of neural network tool used in supervised learning.

Barnes and Laughlin (2002) remark that the step of features' selection is frequently more important than the choice of the classification algorithm.

Geophysicists and geologists can use different techniques to support different stages of the interpretation workflow. For instance, k-means can be suitable for preliminary classification, starting with small K values and gradually increasing the number of class. Then, classification can be refined progressively using other techniques, such as SOMs and/or ANNs.

Finally, multiple learning algorithms can be combined to define a hybrid classification approach. Alternatively, just the outputs of different classification algorithms can be combined. The final objective of these hybrid approaches is to enhance all the strengths and to minimize the weaknesses of each individual algorithm.

5.3.1.2 Musical Domain

Brain imaging techniques show that when we listen to or play music, the entire brain is active (an interesting discussion about the musical instinct in humans, including the cognitive aspects of music, can be found in Ball, 2010). Over the past two decades, many studies confirmed that music can produce anatomical variations in our brain. Schacher and Neff (2015) recently discussed the impact of musical practice on high-level cognitive functions, such as integrating multisensory perception and attention. They showed that improved cognition is the effect of structural changes in the brain.

"Compared to non-musicians ... musicians usually exhibit: a larger and denser corpus callosum connecting the two brain hemispheres, which is developed during the critical phase of juvenile training increasing interhemispheric connectivity and communication; a larger auditory cortex as well as a specific inter-hemispheric asymmetry; a larger and more dense motor cortex, especially in somatotopic representation of the respective extremities (e.g., areas representing fingers in piano players); changes in the cerebellum (responsible for motor control, integration and simulation); increased hard-wired connections between auditory and (pre-) motor areas via fiber bundles (e.g., arcuate fasciculus). Taken together, extensive training followed by expertise — as they are reflected in their neuroplastic correlates — represents the underlying cause for the structural differences between experts and non-experts" (Schacher and Neff, 2015).

Why music have so big effects on our brain? Alternatively, reversing the question, why our brain is so sensitive to music? A possible explanation is that music represents much more than a mere sequence of sounds. Indeed, *a musical piece is formed by structured patterns of sounds.* Our instinctive ability to recognize sound patterns offers obvious adaptive benefits (Huron, 1991). It derives from natural evolution of humans' cognitive capabilities (Ball, 2010). This innate attitude is enhanced if music is integral part of the social/cultural background. Nowadays, human sensitivity for music is supported by advanced technology and by many formats and informatics protocols used in digital music [such as WAV, MP3, and musical instrument digital interface (MIDI)[6]]. These allow easy musical production, sound processing, analysis, integration, automatic pattern recognition, and classification. For instance, Shazam is a popular application for musical pattern recognition. Initially created for music identification, its functionalities have been quickly expanded to other types of media, including cinema and TV. Using a smartphone, it is possible to gather a brief sample of audio for creating an acoustic fingerprint. This is compared against a central database. If it finds a match, information about the artist, song title, and album is sent back to the user.

Algorithms capable of performing automatic recognition and classification of musical pieces are becoming increasingly useful due to rapidly growing networked music archives. Individual users, music librarians, and database administrators use these algorithms to explore huge databases on the web and sorting in different genres their music collections automatically. The science of retrieving information from music represents an emerging set of disciplines known as music information retrieval. It finds the main applications in instrument recognition, automatic music transcription, automatic categorization, and genre classification.

The algorithms of pattern recognition and data mining used in music and geophysics are based on similar principles. Indeed, the classification problem is analogous: exploring a big database and clustering objects with similar features. These features correspond to geophysical or musical attributes, depending on the domain. In both cases these attributes derive mainly (but not exclusively) from basic physical properties, such as the frequency spectrum of the data. High-level features can be added in music. For instance, rhythmic features represent important attributes used in musical pattern recognition and genre classification.

"Musical genre" is one of the most important means for classifying and organizing music. This concept can be assimilated to the concept of seismic facies in geophysics. In both cases, genres and facies represent "classes of objects" that can be clustered based on common features. Musical genres and seismic facies represent fuzzy concepts, rather than unambiguous categories. Determining which musical feature to consider for genre classification is equally important as determining which geophysical attribute to use for facies clustering.

---

[6] MIDI means music instrument digital interface. It is a standard protocol/format used in digital music.

As I said, both classification problems are commonly faced through similar approaches. Supervised and unsupervised learning algorithms are applied for musical classification too. In the first case, machine-learning techniques are used to train on model examples. Previously unseen musical recordings can be classified into one of the genre categories using the rules generated during training. Instead, unsupervised learning algorithms cluster the musical recordings based on similarities rather than model categories.

The classification workflow in the musical domain follows similar stages as in geophysics. The first step is to define genre taxonomy: based on his/her experience, the user decides how many and what musical genres should be considered for classifying his/her database. That a prior choice will be necessarily subjective, but it can be supported by a robust musical background. The second step is selecting the musical features to be considered. Some authors (McKay, 2004) subdivide the musical attributes in three main types: low-level features (spectral or time-domain information extracted directly from audio signals); high-level features (for instance, instruments, melodic contour, chord frequencies, and rhythmic density); cultural features (sociocultural information outside the scope of musical content itself). The second and third types of features offer an advantage in musical genre classification with respect to geophysical facies classification. In particular, rhythmic, melodic, and harmonic properties can be determinant in the process of categorization; they can be easily extracted from musical files recorded in symbolic formats such as MIDI

After having selected and extracted the features, these can be grouped together into *feature vectors*. These will be used as the input to classification systems. A process of training is commonly performed on a subset of the musical recordings. Training can be performed using different feature vectors to verify which combinations perform best.

The classification of the musical data can be finally performed using the selected features and applying one of the many classification methods available. In the same way as it happens in geophysics, there are many methods available for automatically classifying musical recordings based on features. McKay, (2004) provides a good overview and a detailed discussion on some supervised learning methods. The same author discusses a comparative analysis of neural network and k-NN classifiers, showing benefits and limitations of both methods when applied to MIDI files. McKay shows how the classification of this type of files can be efficiently performed using different features including pitch (linked to frequency), velocity (related to sound intensity), melody, harmony, chords, musical texture, and so forth. The final validation phase of the process of classification is performed comparing the results obtained using different types of classification methods and variable feature vectors.

5.3.1.3 *Links*

Geophysical signals are commonly represented as a space/time series of physical events that are characterized by a certain frequency spectrum. Think for instance of a seismic trace. It is the result of a pressure wave produced by an earthquake or by an artificial source in a geophysical survey. It propagates through the Earth, is reflected at geological interfaces and, finally, is recorded by geophones.

From a physical point of view, a musical sound can be seen in a similar way: it is produced by a pressure wave generated by some musical instruments. That wave is characterized by a certain frequency spectrum, propagates through the air, and arrives to our ears. Of course, there are also substantial differences between a seismic recording and a beautiful musical piece. However, it is well known that geophysical signals and sounds can be analyzed using similar approaches, after proper format transformation.

Fig. 5.2 shows some key aspects about the link between seismic data and sounds. The top panel displays a piece of a seismic trace. The central panel is the correspondent spectrogram; it shows the frequency content of the signal versus time. The bottom panel is the correspondent MIDI file derived from the spectrogram (in Chapter 7, I will discuss how to perform the transformation from seismic to musical files). The "MIDI piano roll display" is just a discretized version of the spectrogram itself in terms of digital notes. This MIDI file allows capturing the "musical" nature of the seismic waves, after they have been transposed into the audible range (between about 30 Hz and 20 KHz). The music associated with the seismic signal changes in correspondence of relevant variations in the elastic parameters. The Earth behaves like a sort of musical filter that produces specific sounds in correspondence of variations of elastic parameters. For instance, interesting sound patterns can be captured in correspondence of the reservoir time interval (see the zoomed image).

Converting geophysical signals into sounds is a relatively old idea (Benioff, 1953; Speeth, 1961). It is possible to convert geophysical data into sounds with high accuracy using the proper mathematical transformations. This conversion from the geophysical to the musical domain can introduce many benefits. The first advantage is that coupling the sounds extracted from geophysical data (such as seismic reflections) with the conventional imaging techniques (such as a seismic section) can contribute to describing and to analyzing the physical aspects of the phenomenon itself. Listening to the pitch and volume in synchrony with visual variations of frequency and amplitude increases significantly the amount of information perceived at the same time. The natural attitude of the human brain for music and sound perception, recognition, analysis, and interpretation can be used in geophysics. In other words, two senses

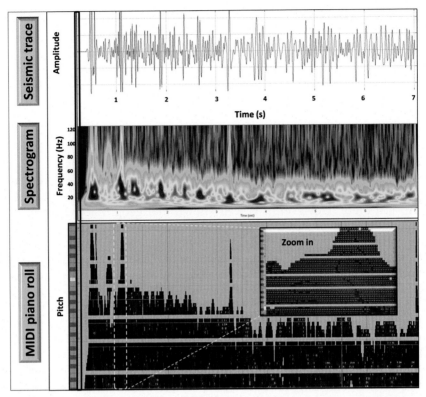

FIGURE 5.2   Example of audio–video display of a seismic trace (top panel) transformed into a MIDI file (bottom panel), derived from its spectrogram (middle panel). Zoom in the reservoir time interval. Warm colors indicate high amplitude and high sound intensity. *MIDI*, musical instrument digital interface.

work better than one. That is true in the ordinary life and it is reasonable to assume analogous benefits also in geosciences.

There is an additional advantage in transforming geophysical data into sounds. After proper format translation (for instance, from the typical seismic format, SEGY developed by the Society of Exploration Geophysicists, to the commonly used musical protocol, MIDI) we can import advanced technology routinely applied in digital music into the geophysical domain. For instance, it is possible to analyze big seismic volumes with the same powerful algorithms of clustering commonly applied in musical genre classification. We can apply in geophysics similar pattern recognition approaches commonly used for searching musical pieces in wide libraries available on the web. These algorithms, properly adapted, can be addressed to geological and geophysical purposes, such as facies classification, formation evaluation, reservoir characterization, and so forth. In the part of this book dedicated to "brain-based technology," I will expand these concepts showing some real applications in hydrocarbon exploration.

Finally, seismic patterns can be transformed into MIDI patterns and analyzed from a different "musical" perspective (Fig. 5.3). This idea has been effectively developed and successfully tested on real seismic data sets (Dell'Aversana, 2013b, 2014). It will be further discussed in the Chapter 7.

5.4 SUMMARY AND FINAL REMARKS

Recognition is the base of cognition. The mammalian brain is structured for allowing quick recognition of "objects" that are crucial for

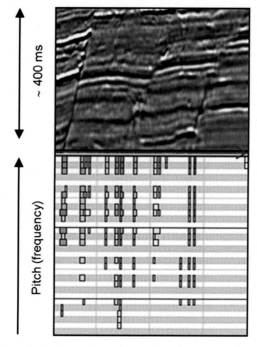

FIGURE 5.3   Portion of seismic section with faults (upper panel) and corresponding musical instrument digital interface (MIDI) file extracted at the depth indicated by the *red arrow* (lower panel). The pitch of the MIDI events shown in the lower panel has been transposed toward high-frequency octaves (>500 Hz) to highlight their musical effect. In this color scale, blue means high intensity and white means low intensity. Comparing the two panels, large faults on the seismic image correspond with clear MIDI events. Moreover, there are additional MIDI events caused by minor lateral discontinuities that are not visible or not clear on the seismic imaging. *After Dell'Aversana, P., 2014. A bridge between geophysics and digital music. Applications to hydrocarbon exploration. First Break 32 (5), 51–56.*

survival. Moreover, the neocortex evolved in humans for refining our pattern recognition capabilities. Nowadays, we continuously perform processes of pattern recognition and classification at extremely variable level of complexity. These processes range from immediate and instinctive recognition of familiar faces, songs, and images to classification of complex information extracted from big data. In human beings, pattern recognition is the result of the continuous interplay between innate capabilities and rigorous scientific approaches. Nowadays, the instinctive ability to recognize natural shapes is supported by automatic pattern recognition algorithms: this hybrid approach represents the norm in the work of many geoscientists. Geophysicists who interpret seismic data commonly use both the instinctual and the technological means for recognizing key features in the data.

As explained in the previous chapter, our brain works creating maps and images of the external and internal worlds. We not only map images but also sounds, movements, internal perceptions, emotions, and expectations. Consequently, the process of pattern recognition concerns our multisensory perception and multimodal cognition. This fundamental concept represents the cognitive background for developing new technologies of data interpretation, mining, clustering, and classification. These new approaches use multimodal pattern recognition of images and sounds based on multilevel feature extraction. After proper transformation of geophysical data into formats used in digital music, new features can be introduced for enhancing the process of pattern recognition. This idea has been successfully tested on real seismic data sets and will be discussed in the part of the book dedicated to brain-based technologies.

References

Ball, P., 2010. The Music Instinct: How Music Works and Why We Can't Do without it. Italian Translation. Edizioni Dedalo srl, 2011.
Barnes, A.E., Laughlin, K.J., 2002. Investigation of methods for unsupervised classification of seismic data. In: 72nd Annual International Meeting. SEG, pp. 2221–2224. Expanded Abstracts.
Benioff, H., 1953. Earthquakes around the world. In: Cook, E. (Ed.), On Out of This World. Side 2.
Bishop, C.M., 2006. Pattern Recognition and Machine Learning. Springer.
Dell'Aversana, P., 2014. A bridge between geophysics and digital music. Applications to hydrocarbon exploration. First Break 32 (5), 51–56.
Dell'Aversana, P., 2013a. Cognition in Geosciences: The Feeding Loop Between Geo-disciplines, Cognitive Sciences and Epistemology. EAGE Publications, Elsevier.
Dell'Aversana, P., 2013b. Listening to geophysics: audio processing tools for geophysical data analysis and interpretation. The Lead. Edge 32 (8), 980–987. http://dx.doi.org/10.1190/tle32080980.1.
Duda, R.O., Hart, P.E., Stork, D.G., 2001. Pattern Classification, second ed. John Wiley & Sons.

Edelman, G.M., 1987. Neural Darwinism: The Theory of Neuronal Group Selection. Basic Books, New York, ISBN 0-19-286089-5.

Edelman, G.M., 1992. Bright Air, Brilliant Fire: On the Matter of the Mind. Reprint Edition 1993. Basic Books, ISBN 0-465-00764-3.

Gilbert, P.E., Kesner, R.P., DeCoteau, W.E., 1998. Memory for spatial location: role of the hippocampus in mediating spatial pattern separation. J. Neurosci. 18, 804–810.

Huron, D., 1991. The avoidance of part-crossing in polyphonic music: perceptual evidence and musical practice. Music Percept. 9, 93–104.

Kiang, M.Y., 2003. A comparative assessment of classification methods. Decis. Support Syst. 35, 441–454.

Kitzbichler, M.G., Smith, M.L., Christensen, S.R., Bullmore, E., 2009. Broadband criticality of human brain network synchronization. PLoS Comput. Biol. 5 (3), e1000314.

Kurzweil, R., 2012. How to Create a Mind: The Secret of Human Thought Revealed. Viking Books, New York, ISBN 978-0-670-02529-9.

Mandelbrot, B., 1967. How long is the coast of Britain? Statistical self-similarity and fractional dimension. Science 156, 636–638.

Mandelbrot, B., 1982. The Fractal Geometry of Nature. Freeman.

Mattson, M.P., 2014. Superior pattern processing is the essence of the evolved human brain. Front. Neurosci. 8, 265. http://dx.doi.org/10.3389/fnins.2014.00265. Published Online 2014 Aug 22.

McKay, C., 2004. Automatic Genre Classification of MIDI Recordings. Thesis Submitted June 2004. Music Technology Area Department of Theory, Faculty of Music, McGill University, Montreal.

Pearce, J.M., Roberts, A.D., Good, M., 1998. Hippocampal lesions disrupt navigation based on cognitive maps but not heading vectors. Nature 396, 75–77. http://dx.doi.org/10.1038/23941.

Schacher, J.C., Neff, P., 2015. The fluid and the crystalline − processes of the music performing and perceiving body. In: Proc. of the 11th International Symposium on CMMR, Plymouth, UK, June 16–19, 2015.

Speeth, S.D., 1961. Seismometer sounds. J. Acoust. Soc. Am. 33, 909–916.

Spiers, H.J., Burgess, N., Hartley, T., Vargha-Khadem, F., O'Keefe, J., 2001. Bilateral hippocampal pathology impairs topographical and episodic memory but not visual pattern matching. Hippocampus 11, 715–725. http://dx.doi.org/10.1002/hipo.1087.

Turcotte, D.L., 1997. Fractals and Chaos in Geology and Geophysics. Cambridge University Press, Cambridge, ISBN 0-521-56733-5.

Zhao, T., Jayaram, V., Roy, A., Marfurt1, K.J., 2015. A comparison of classification techniques for seismic facies recognition. Interpretation 3 (4), SAE29–SAE58.

Web References

Geoteric website: http://www.geoteric.com/geoteric-CI.

Further Reading

Turcotte, D.L., Schubert, G., 2002. Geodynamics, second ed. Cambridge University Press, ISBN 0521666244.

Integration

6.1 GEOPHYSICAL PROBLEMS

Our brain continuously combines heterogeneous information into a coherent perception. In most cases, this process happens unconsciously during our daily routine. In the scientific domain, integration of complementary data is, or should be, driven by an intentional and rational approach. In the Earth disciplines, data are integrated at many different spatial and temporal scales and for variable purposes. For instance, in the hydrocarbon industry, there is a growing interest in acquisition and integration of multidisciplinary geophysical data. The recent technological and scientific development supports and encourages the acquisition of huge multidisciplinary data sets. Of course, benefit can be extracted from multidomain, multiscale data sets if and only if integration is performed appropriately (Dell'Aversana, 2014). There is always the risk that redundant data sets can produce chaos instead of knowledge improvement. In fact, an inappropriate integration workflow can lead toward wrong Earth models, degrading geophysical imaging and increasing interpretation ambiguities. For this reason, quantitative integration in geosciences should be considered as a rigorous discipline by itself. Moreover, acquiring, processing, and combining constructively different geophysical information can be very expensive. Consequently, the cost-to-benefits ratio must be evaluated in each specific case.

In summary, from a pragmatic point of view, integration is an advantageous process if it reduces the uncertainties about some properties of interest, supporting decision-making. Furthermore, the costs for combining complementary information must be significantly lower than the economic benefits introduced by the integration process.

A necessary condition for a fruitful process of integration is that the different types of information must be linked in some way. Fig. 6.1 shows schematically that the link between different geophysical domains [seismic, electromagnetic (EM), and gravity, in this example] can arise at

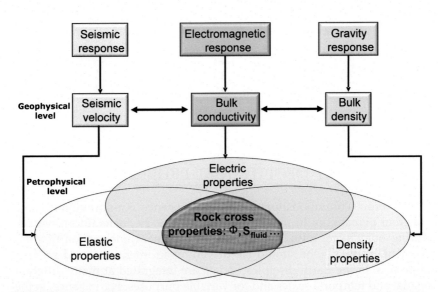

FIGURE 6.1 Conceptual scheme of the links between different geophysical domains. *After Dell'Aversana, P., 2014. Integrated Geophysical Models: Combining Rock Physics with Seismic, Electromagnetic and Gravity Data. EAGE Publications.*

different conceptual levels and with variable physical meaning. The different domains are linked at a *geophysical level* by relationships between seismic velocities, density, and electrical resistivity (generally indicated as geophysical properties or geophysical parameters). Moreover, there is a "deeper" *petrophysical link*. In fact, a common set of physical properties (symbolically represented in gray in Fig. 6.1) influences all the above-mentioned geophysical parameters. These properties concern the rock matrix and the fluids in the rocks, such as porosity (Φ) and fluid saturation (S_{fluid}). They are often indicated as rock cross properties.

The figure suggests that combining seismic, gravity, and EM information can be useful for estimating fluid saturation and porosity with a satisfactory level of reliability. Instead, seismic data and EM (or gravity) observations used independently can only provide an ambiguous estimate of these properties. In fact, it is well known that the seismic response depends significantly on the porosity of the rocks, but it is scarcely sensitive to variable saturation of oil and brine in the rocks. The reason is that the elastic properties are not very different for these two fluids. Instead, the electrical resistivity changes significantly (from one to three orders of magnitude) for water and hydrocarbons and affects strongly the EM response. Thus, integrating seismic and EM data allows estimating porosity and saturation with good accuracy.

The strength of integrated geophysical approaches emerges especially when geological complexity cannot be faced by using a single method. For instance, in hydrocarbon exploration, the geological setting can be extremely complex, as in thrust belts, or in the presence of salt domes or volcanic sequences. In these cases, the use of complementary geophysical approaches can represent the optimal exploration strategy. In addition, in other fields, such as geothermal or mining exploration, the complexity of the geophysical problem can justify an integrated approach. The crucial question is then how to optimally combine heterogeneous information for producing integrated geophysical models.

6.1.1 Integration Methods

Different approaches are used for combining multidisciplinary geophysical data, depending on many factors such as the type of data, the final objective, and the geological setting. In hydrocarbon exploration, the integration workflow is generally different if it is directed to background[1] or to reservoir characterization because the physical properties of interest are generally different in these two cases. In the first case, geoscientists are commonly interested in estimating seismic velocity, resistivity, and

[1] The term "background" is here intended as the spatial geophysical domain in which the reservoir is embedded.

density. Instead, reservoir formations are commonly characterized in terms of fluid saturation, porosity, permeability, pore pressure, elastic constants, and clay mineral content. In other fields of study, integration of geophysical data can be supported also by geochemical information. For instance, combining geophysical, geochemical, and mineralogical data with detailed historical reconstructions is the way for studying many volcanoes and their deep feeding system. This is the case of Mount Vesuvius, one of the most famous examples of well-studied volcanoes using a multidisciplinary approach.

Integration approaches are used in other disciplines too, such as geothermic exploration, civil engineering, environmental geology, archaeological investigations, hydrogeology, and geomorphology. Despite the methodological differences of these disciplines, the approach for combining multiphysical measurements, such as seismic, EM, gravity, and magnetic data, is based on similar criteria.

Geoscientists commonly start with a qualitative approach based on mapping different physical responses and comparing them. Fig. 6.2 shows an example of comparison between gravity and EM responses measured in a large exploration area. A map of the normalized electric field (with frequency of 0.5 Hz) observed at seafloor is superimposed on

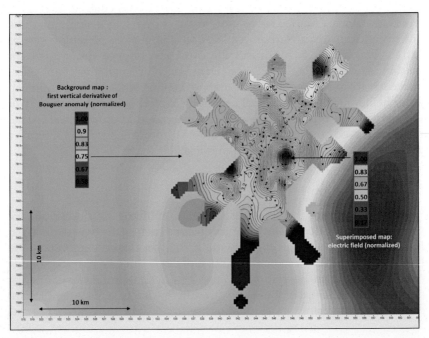

FIGURE 6.2   Corendered maps of electromagnetic anomaly (small map with contour lines) and gravity anomaly (background map).

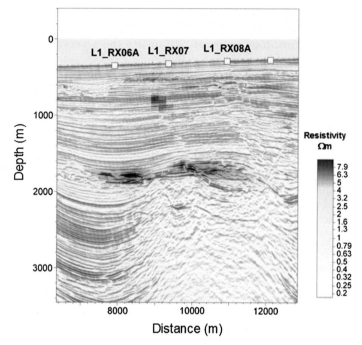

FIGURE 6.3   Example of resistivity model obtained by electromagnetic inversion constrained by seismic reflections and corendered with the seismic section (in background).

the vertical derivative of the Bouguer anomaly.[2] Combining different types of geophysical information by simple superposition or direct comparison is useful for checking reciprocal consistency and possible correlations among different physical responses. Comparing images, models, and maps is a good starting point for supporting a more quantitative integration workflow. This can continue through constrained inversion.[3] This is frequently applied when we deal with two or more data sets characterized by different accuracy. In this technique, the most robust data set is commonly used for constraining or driving the inversion of data characterized by a higher level of uncertainty. For instance, sharp-boundary information, as provided by seismic data, can be used to constrain the inversion of EM or gravity data.

Fig. 6.3 shows an example of resistivity model (in colors) corendered with seismic data. It was obtained through inversion of EM data

[2] The vertical derivative of the Bouguer anomaly works as a high-pass filter and allows highlighting geological details at lower spatial scale.

[3] A constraint is commonly intended to be some type of information in which we have strong confidence. It can also be an assumption or any other a priori information that we intend to keep fixed during the inversion of our experimental data set.

constrained by seismic reflections. These provide the geometry of the top and the bottom of the reservoir layer. In this case, the inversion of the EM data is aimed at estimating resistivity inside the reservoir. The figure shows an anomaly of resistivity (in red) with respect to the background (in white). It is caused by the presence of oil in the geological formation, as confirmed by drilling results (Dell'Aversana et al., 2011).

A more complex approach is based on sequential loops of inversion performed in different geophysical domains: in this technique, the output model of one inversion process is transformed into an initial model that triggers inversion in a different geophysical domain (Dell'Aversana, 2001, 2003). For instance, the seismic velocity model obtained through seismic tomography can be transformed into a resistivity model using some type of empirical relationship or using rock physical models available in the scientific literature. This "derived model" is then used as a starting guess for running the inversion of EM measurements. The loop can be iterated many times, until a satisfactory velocity-resistivity model is obtained fitting all the data in both seismic and EM domains.

The most complex method of integrating different data sets is simultaneous joint inversion (Appendix 2). In this case, a composite (or joint) cost function is minimized. This function includes the misfit between observed and predicted values for two or more independent data sets belonging to different domains. For instance, simultaneous joint inversion of EM and seismic data allows obtaining robust estimation for both electrical and elastodynamic parameters. Appropriate cross-domain relations between different rock properties are used for linking the different geophysical domains. The result is a multiphysics model honoring all the measurements, plus the relationships between the different physical parameters. The uncertainties that affect the inversion results when running single domain procedures are significantly reduced using this joint inversion approach.

When powerful computing resources are available, geophysicists prefer combining different integration approaches in optimized workflow (Dell'Aversana et al., 2016). Modeling and inversion algorithms are merged with an integration architecture that allows managing data and models coming from different domains, with the support of smart visualization tools. This "system of integration" is commonly implemented into a comprehensive software and hardware platform that includes many advanced codes. These work in cooperation using powerful computer clusters. Fig. 2.9 is a good example of multiparametric geophysical models obtained combining many different integration approaches, such as single domain, constrained, cooperative, and joint inversion of seismic, gravity, and EM data.

After creating a multiparametric geophysical model, geoscientists must interpret it geologically. The process of integration continues through the combination of additional information derived from surface geology, well logs, laboratory data, and so forth. The final objective is to define the most reliable geological model that honors all the available observations.

6.2 NEUROBIOLOGICAL BACKGROUND

As remarked by the neuroscientist Giulio Tononi, "The brains of higher mammals are extraordinary integrative devices. Signals from large numbers of functionally specialized groups of neurons distributed over many brain regions are integrated to generate a coherent, multimodal scene. Signals from the environment are integrated with on-going, patterned neural activity that provides them with a meaningful context" (Tononi et al., 1998).

Indeed, integration is a crucial function of the brain of mammals, especially of the human brain. It is not only different parts of the cortex that are reciprocally integrated through complex connections between neural groups (Edelman, 1992). Subcortical nuclei also play a fundamental role in integration, like the thalamus, colliculi, and other neural structures located in primordial areas of the brain (see Fig. 2.2).

The ability of our brain to combine constructively streams of heterogeneous information represents a significant adaptive advantage. Real-time integration of multiple sensorial stimuli into coherent concepts speeds up the decisional process and optimizes the interaction with the environment. Integration of brain functions is fundamental for an additional reason. Indeed, the neuroscientists Edelman and Tononi have widely recognized and discussed the importance of integration for explaining the most crucial aspect of high-level human cognition, consciousness.

6.2.1 Integration and Consciousness

6.2.1.1 Edelman's Theory

The theory of selection of neural groups proposed by Gerald M. Edelman is based on his idea of neural Darwinism (Edelman, 1987). The brain is a selective recognition system. It evolved as the biological response of organisms to external inputs from the environments. That neurobiological system is able to cluster sparse information for transforming it into coherent concepts. Since the birth, the brain of mammals is already provided with a *primary repertory* of synaptic connections. These are continuously modified depending on the environmental inputs. Only those connections, which allow producing some advantageous behavior, are enhanced and stabilized. Daily experience is responsible for developing a *secondary repertory* of connections between neurons. These connections form a set of neuronal groups that work (oscillate) in synchrony. Moreover, the groups are progressively linked to each other and form "super groups" working as cognitive units of higher level.

There are three main aspects of this theory. The first is that the neural selection derives from experience: only some connection predominates in response to the stimuli from the environment. The second crucial aspect is recognition. This is intended by Edelman as the continuous fit between the elements of a physical domain (the brain) with the events happening in another physical domain (the environment). The brain is intrinsically linked with the ecosystem and changes in response to events happening in both the environment and the body.

Finally, the third fundamental aspect is integration. This is biologically expressed in terms of connections between elements characterized by a specific individuality (single neurons) but interdependent at the same time (they are activated in a synchronic way). The brain changes through a continuous process of integration that allows the formation of aggregates at variable levels of complexity. These aggregates represent the biological background of cognitive functions, such as perceptive categorization, memory, and learning. Edelman remarks that the whole process is self-organized, and it does not need any supervising algorithm somewhere in the brain. It does not require either top-down instructions or a center of control and coordination.

Edelman explains the intelligent behavior as global cognitive feature emerging from the interaction of the brain, body, and environment (Edelman, 1992). Consciousness can be explained as a property emerging from the dynamics of this complex interaction. The majority of the superior mammals have only a "primary consciousness." This is the capability to perform complex discrimination of the events happening in their experience but without being conscious of that process. Only humans are able to be conscious of being conscious (higher-order consciousness). This superior cognitive ability comprises an awareness of the past, the future, and the *self* that is aware of them. It requires a powerful memory system and the language. This provides the brain with the repertory of symbols necessary for building complex concepts.

6.2.1.2 *Tononi's Theory*

Giulio Tononi et al. proposed a theory (integrated information theory, briefly IIT) for explaining how the process of integration of information can allow the development of consciousness (Tononi, 2010; Oizumi et al., 2014). These authors formulated the fascinating hypothesis that consciousness could be a property of all informative systems that are characterized by a high level of complexity in terms of connections between their constituent parts.

Two main factors are responsible for generating consciousness: the coexistence of functional specialization and their integration in networks.[4] These two conditions arise in the mammalian brains. In humans, virtually any neuron in the corticothalamic system can potentially interact with any other neuron, creating an unpaired informative connectivity (about 10^{15} neural connections are estimated in our brain).

Following Tononi, the quantity of consciousness is determined by the amount of integrated information, whereas the set of informational relationships determine the *quality of consciousness* (Tononi, 2010).

I would like to remark that the theory proposed by Tononi is fully consistent with my "generalized theory of integrated information" (Dell'Aversana, 2013) based on the concept of "semantic entropy." This is a nonlinear function of the ratio between two terms: the total amount of information available by the informative system and the amount of linked (integrated[5]) information. The informative system can be an individual brain but also an entire community of brains.[6] These systems continuously work for linking heterogeneous streams of data to decrease semantic entropy. When the process of integration is successful, the result is a progressive increase of structured information, with formation of robust relationships and data correlations. The informative complexity is transformed into organized semantic structures, such as significances, concepts, models of the reality.

In summary, from a neurobiological point of view, integration of information corresponds to integration of neurons and connections between different brain regions. This process happens at variable levels of complexity and at different spatial and temporal scales. It can be confined to a restricted number of local synapses or can be extended to links between distant neural maps. The different levels of cognition derive from the ability of the brain to create neural connections at variable level of complexity. These allow organizing the continuous information streams coming from the external and the internal worlds, allowing our mind to have a unitary perception of the world.

---

[4] Of course, these must be intended as necessary but not sufficient conditions. As well remarked by Edelman, also other fundamental conditions are required for creating high-order consciousness, such as a long-term memory and the availability of a cognitive system able to manage symbolic information. Furthermore, Damasio, Panksepp, and other neuroscientists, remarked the role of the so-called "emotional brain" that includes ancient subcortical nuclei.

[5] How the quantity and the quality of integration affect the value of semantic entropy is explained with additional details in my previous book about cognitive aspects of geosciences (Dell'Aversana, 2013).

[6] These are called "Semantic Systems" (Dell'Aversana, 2013) because they intentionally create significances through integration of heterogeneous streams of information.

6.3 LINKS AND EXAMPLES

Integrating heterogeneous data is a common task in many interpretative disciplines including Earth and medical sciences. The main reason is that a geological model, as well as a medical diagnosis, represents the result of an interpretation activity rather than the output of a deterministic workflow. Consequently, it is reasonable to support the analysis combining complementary information to reduce the unavoidable uncertainties. For instance, integration of seismic and nonseismic data is crucial in complex exploration areas (Dell'Aversana, 2014). Analogous benefits are offered by integrated approaches applied in medical disciplines, where hybrid imaging methods demonstrate their effectiveness in solving many problems. This approach is also known as "medical image fusion." This is the process of retrieving crucial information from multi-modality medical images for various purposes, such as tumor detection, diagnosing diseases, surgery treatment, and so forth. Independently from the field of application, combining complementary data and methods commonly minimizes the intrinsic limitations and enhances the advantages of each individual approach.

However, integration is generally a costly and a risky process. It commonly requires allocating supplementary budget and resources in terms of human expertise and technology. Moreover, integration is often a time-consuming process that not necessarily improves the results. For instance, in geosciences, including EM data can generate confusion and conflicts with physical models based on seismic data. Sometimes EM and seismic responses can appear to be reciprocally inconsistent: it can happen that significant seismic anomalies are detected at the same location where the EM and/or gravity responses appear homogeneous (Dell'Aversana and Vivier, 2009).

These conflicts of interpretation are frequent in hydrocarbon exploration because the effect of oil and gas in the rocks can be sensed with different sensitivity by EM, gravity, and seismic methods.[7] An analogous problem arises also in medical diagnosis. Sometimes it happens that increasing the number of analysis generates additional ambiguity about the diagnosis itself, rather than reducing the uncertainty.

---

[7] Uncorrelated geophysical responses observed at surface should be explained in any case. In fact, they can provide deep insight about hydrocarbon distribution in the geological formation. For instance, a few percentage of gas saturation in sandy formations can be sufficient for creating a significant seismic anomaly without any detectable electromagnetic response. Thus, comparing the two types of seismic and EM data can be useful for distinguishing commercial reservoirs from noncommercial gas accumulations. Apparent inconsistencies between different geophysical responses should be considered as useful information, rather than misleading.

In summary, combining different types of information and solving apparent contradictions require significant additional efforts. Consequently, a crucial question is how to estimate the cost-to-benefit ratio of integration. In practice, allocating additional resources for acquiring and integrating complementary data should be justified only if some additional value is expected. Thus, the question becomes how to estimate the value of integration (VOINT). A general criterion is to evaluate how the additional data can modify, in a probabilistic sense, our actual knowledge, affecting our decisions. The starting point of this process of evaluation is to estimate the so-called "value of information" (VOI).

6.3.1 The Value of Information

The brain continuously combines new data with a prior knowledge. When our senses perceive new information, this is combined into a personal *system of knowledge*, including previous data saved in our memory, subjective sensations, expectations about the future, beliefs, hopes, and so forth. Our mind continuously performs a complex, often unconscious, work of integration between heterogeneous data belonging to the past, present, and future time. Consequently, our decisions and behaviors are affected by the new information just perceived combined with our prior knowledge. An analogous situation arises in many experimental sciences. When a new set of observations O is acquired, it is commonly combined with prior knowledge. Consequently, our current theories, hypotheses, and expectations are eventually updated by the new information. Let us suppose that we use our current knowledge and experience for defining the prior probability of a certain scenario H. For instance, the geophysical information in our current database supports the scenario ($H = oil$) of a possible hydrocarbon discovery at a given location, with a certain prior probability of success, $P(oil)$.

If we acquire a new set of geophysical data, O, this can change the prior probability of the scenario H into a new *conditioned probability*. That conditioned probability is commonly indicated with the notation $P(oil|O)$. It indicates the posterior probability of having an *oil scenario*, given the set of observations O. The question is "what is the new value of this posterior conditioned probability?" This problem is commonly formalized using the Bayes formula, well known in statistical applications. It provides the conditioned (posterior) probability $P(H|O)$ for a scenario H after modification of unconditioned probability (a priori probability) due to a set of observations O:

$$P(H|O) = \frac{P(O|H) \cdot P(H)}{P(O)}. \qquad (6.1)$$

Symbols in Formula (6.1) have the following meaning:

1. $P(H)$ is the a priori probability (without observations O) that the scenario H is verified;
2. $P(O|H)$ is the probability of the observations O when the scenario H is verified;
3. $P(O)$ is a normalization factor that assures condition $0 \leq P(H|O) \leq 1$. Its general form is given by:

$$P(O) = \sum_{i=1}^{n} P(O|H_i) \cdot P(H_i), \tag{6.2}$$

where H_i is the generic scenario (*i*th) in a set of n scenarios.

The Formula (6.1) is often indicated as *Bayes theorem*. Its demonstration is immediate (Russell and Norvig, 2003); however, the formula is very intuitive. It says that the product of two probabilities (the denominator is just a normalization factor) gives the posterior probability. One is the prior probability of the scenario H. For instance, it can be statistically estimated. The other one is the conditioned probability of observing the data set O if the scenario H is verified. For instance, this can be estimated through modeling (simulating the same type of measurements in the scenario H).

Returning to our geophysical examples, for instance, we could be interested in estimating how a set of EM measurements [controlled source electromagnetic (CSEM)] can affect our probabilistic estimation of doing a commercial hydrocarbon discovery in a certain exploration block. CSEM data frequently add useful information, thus we expect that the new data set can improve our knowledge, reducing the exploration risk in that area.

In this example, H_i in Formula (6.2) can be the scenario of a commercial oil discovery or the scenario of a dry well. For sake of simplicity, we can start defining the information O as a binary indicator, where O is a *positive indicator* if it increases the chance of success (for instance, a commercial discovery) and a *negative indicator* otherwise.[8] Observation O can be a CSEM data set, but the same discussion can be done for any other type of information. For instance, we can consider the following case:

1. $P(H) = P(oil)$ is the unconditioned (a priori) probability of drilling an oil-filled reservoir. This probability can be statistically estimated based on previous exploration results in the same area, on geological studies, and so forth.

---

[8] Following the well-known medical definitions, positive and negative indicators can be further divided into "true positive," "false positive," "true negative," and "false negative." For instance, a CSEM anomaly will be a true positive hydrocarbon indicator if it corresponds to a true hydrocarbon discovery. Otherwise, the same CSEM anomaly will be a false positive if it corresponds to a dry well.

2. $P(O|H) = P(\text{CSEM}/oil)$ is the conditioned (a posteriori) probability of having a significant CSEM anomaly (for instance, normalized electric magnitude >20%), if a commercial oil reservoir is effectively present at that location.

For a fixed scenario H (*oil, no oil* ...), the set of observations O (CSEM anomaly, no CSEM anomaly ...) can give a true positive or a false positive. For instance, in case we observe a significant CSEM anomaly, the posterior probability for an oil scenario is given by Bayes formula:

$$P(oil|\text{CSEM}) = \frac{P(\text{CSEM}|oil) \cdot P(oil)}{P(\text{CSEM}|oil) \cdot P(oil) + P(\text{CSEM}|no\ oil) \cdot P(no\ oil)}. \quad (6.3)$$

The denominator in Formula (6.3) acts as a normalization factor, to have $0 \leq P(oil|\text{CSEM}) \leq 1$. In this specific example, it represents the sum of the probability to have a true positive and the probability to have a false positive. An analogous formula can be written in case of no detection of CSEM anomaly.

Buland et al. (2011) estimated the impact of CSEM on risk analysis using this Bayesian approach. They calculated the risk modification applying the Bayes theorem when EM information is taken into account. The authors confirmed that CSEM method can provide useful information improving the process of risk evaluation in exploration, increasing significantly the confidence of discovering a commercial hydrocarbon reservoir.

The value of CSEM information (VOI) depends on how it affects the exploration decisions. If the drill cost is C, the net present prospect value is V (excluding drilling cost), and a prior chance of success (without CSEM anomaly) is $P(oil)$, then the expected net present value without CSEM information is (Buland et al., 2011):

$$E(V) = P(oil) \cdot V. \quad (6.4)$$

Instead the expected net present value with CSEM information is

$$E(V)' = P(oil|\text{CSEM}) \cdot V, \quad (6.5)$$

where $P(oil|\text{CSEM})$ is provided by Formula (6.3).

The VOI of CSEM is given by the estimated value of the prospect with CSEM information (Formula 6.5) minus the estimated value of the prospect without CSEM information (Formula 6.4), minus the cost C of the additional information.

6.3.2 The Value of Integration in Geophysics

To highlight the importance of integration of multidisciplinary data sets, we can expand the same Bayes theorem including additional types of geophysical information. The objective is to decrease the exploration risk. I assume that adding complementary geophysical data can increase the probability of an exploration success.[9] Of course, this additional data set has a cost. Thus, we need to verify if the *additional VOINT* of independent data is greater than the *additional cost* for acquiring, processing, interpreting, and integrating the complementary data.

As demonstrated in the previous works, gravity information can change the interpretation of ambiguous CSEM response (Dell'aversana et al. 2012). The key concept is that the degree of spatial correlation between gravity and CSEM anomalies can be used as a qualitative indication for interpreting the geological nature of both gravity and EM responses. In fact, both methods are sensitive to large-scale geological variations. These are generally associated with correlated changes in density and resistivity distribution. For instance, in correspondence of the boundaries of a carbonate formation characterized by high resistivity and high density, we expect to observe high value of both electric and Bouguer anomalies. In that case, we observe correlated CSEM and gravity responses at surface. On the other side, the gravity and the CSEM methods have different spatial resolution. In case of relatively small changes at reservoir scale, for instance, caused by variations in fluids, we expect to observe only CSEM effects at surface. The gravity method has not sufficient resolution for being sensitive to such a small fluid change. Consequently, in that case, we can see uncorrelated CSEM and gravity responses. Therefore, the level of correlation (or uncorrelation) between CSEM and gravity responses can provide some useful indication for deciding if an EM anomaly is more probably caused by presence of hydrocarbons or by resistive and dense geological formations.

Fig. 6.2 shows an example of that concept. It shows a map of the normalized electric field (for a frequency of 0.5 Hz) observed at seafloor superimposed on a high-pass filter of the Bouguer anomaly. We can observe several areas where the gravity and the CSEM anomalies are strongly uncorrelated. The reason is that the two methods (gravity and CSEM) sense different causal bodies at different spatial scale. The gravity method senses mainly the geological trend, such as the variable depth of the carbonate basement. This is relatively shallow at east and produces a strong Bouguer anomaly, as showed in the figure. Instead, the EM anomaly has enough resolution to sense also variations in fluid distribution at reservoir scale. In other words, a possible[10] hydrocarbon-filled

---

[9] This assumption is fully justified by many successful exploration case histories based on integrated geophysical approaches (Dell'Aversana, 2014).

[10] I remark that all these considerations have only a probabilistic value. This value is commonly based on interpretative criteria, generally based on previous geological knowledge in the explored area.

reservoir is sensed by the CSEM method and not by gravity. Consequently, we observe uncorrelated Bouguer and CSEM responses. For instance, a strong EM anomaly, not correlated with the gravity trend, appears in the center of the map (in red), where the hydrocarbon reservoir has been confirmed by several wells.

In different geological scenarios, strong correlation between gravity and EM anomalies has been interpreted in terms of geological transitions from *low resistivity—low density* formations to *high resistivity—high density* formations (Dell'Aversana, 2001, 2003).

In summary, the level of correlation between gravity and CSEM responses can be used for interpreting EM anomalies. The idea is to use combined CSEM and gravity maps for separating anomalies having a probable geological cause (here called "geological scenario") from anomalies interpretable as hydrocarbon indicators (here called "hydrocarbon scenario"). This can be done applying a similar Bayes formula as Eq. (6.1) but slightly more complicate.

To estimate the posterior probability for a given scenario H using both CSEM and gravity observations (and eventually, seismic, borehole, and so forth), we need to apply the Bayes formula adapted for combining different types of observations O_1, O_2, ..., O_n (Russell and Norvig, 2003). This problem can be simplified if we assume that CSEM and gravity anomalies represent independent observations. In fact, even though they can have a common cause, they are acquired independently and there is not any mutual influence (hypothesis of conditional independency). In that case, we can calculate the posterior probability of having, for instance, a geological scenario observing both a significant CSEM anomaly and a significant gravity anomaly at the same location (case of high spatial correlation). It is given by:

$$P(\text{geol}|\text{CSEM and GRAV}) = \frac{P(\text{CSEM}|\text{geol}) \cdot P(\text{GRAV}|\text{geol}) \cdot P(\text{geol})}{P(O)},$$

(6.6)

where $P(\text{geol})$ is the a priori probability to have "significant" geological variations at reservoir depth. It can be estimated through previous experience in the same area or using a geological model. $P(\text{CSEM}|\text{geol})$ is the conditioned probability of having a significant CSEM anomaly (e.g., normalized value >20%) in correspondence of a significant geological change. It can be estimated by modeling. $P(\text{GRAV}|\text{geol})$ is the conditioned probability of having a significant gravity anomaly in correspondence of a significant geological variations. It can be estimated by modeling. $P(O)$ is the normalization factor for assuring that $0 \leq P(\text{geol}|\text{CSEM and GRAV}) \leq 1$.

An analogous formula can be written for estimating the following a posteriori probability:

$$P(oil|\text{CSEM and NO GRAV}) = \frac{P(\text{CSEM}|oil) \cdot P(\text{NO GRAV}|oil) \cdot P(oil)}{P(O)}.$$

$$(6.7)$$

This formula gives the conditioned (posterior) probability to have an oil reservoir after observing a significant CSEM anomaly in absence of any significant gravity anomaly (NO GRAV). In other words, Eq. (6.7) allows estimating the probability of an "oil scenario" based on scarce or null spatial correlation between CSEM and gravity anomalies (such as in the proven oil discovery related to Fig. 6.2).

Example

As a simple exercise, I applied this Bayesian approach for estimating the economic benefits induced by integration on drilling operations in a real exploration area (Fig. 6.2).

First, I assumed that the value of a well could be estimated in terms of drilling cost. In this case, the drilling operations are very expensive. I estimated a prior drilling cost of 70 million dollars with an a priori chance of success of 57%. This probability is based on previous exploration results obtained using seismic data and geological knowledge of the area.[11] Multiplying these values, we obtain about 40 million dollars: that is the *expected risked value* of one well without CSEM and gravity data (but using previous knowledge, including seismic data).

Second, I estimated, through the Bayesian approach previously explained, the posterior chance of success after acquiring gravity and CSEM data and after integrating them: it increased to 86% after combining the CSEM and gravity maps. The value of 86% comes out from Eqs. (6.6) and (6.7) that allow estimating the probability of a "geological scenario" and of an "oil scenario," respectively, based on low spatial correlation between CSEM and gravity responses. Simply speaking, my confidence in CSEM data as possible hydrocarbon indicators is high in this area because there are significant CSEM anomalies uncorrelated with the gravity response (see Fig. 6.2). To quantify my confidence, I performed gravity/CSEM modeling in several realistic geological scenarios constrained by seismic data interpretation. Finally, I transformed the modeling results into probability values[12] to be used in Eqs. (6.6) and (6.7).

---

[11] Of course, these values are indicative: the same Bayesian approach can be applied using different cost scenarios and values, without changing the methodological aspects.

[12] This type of iterative simulation allows estimating, for instance, the probability $P(\text{CSEM}|oil)$ of observing a significant CSEM anomaly, like the red one in the center of the map of Fig. 6.2, in case of a hydrocarbon-filled layer at reservoir depth.

If we multiply the prior drilling cost (70 million dollars = a prior value of the well) by the posterior chance of success (0.86), we obtain about 60 million dollars. This is an estimation of the *expected risked value* of the well *after including CSEM and gravity information* into the risk analysis workflow. Assuming a total cost of 5 million dollars for CSEM and gravity surveys, the total added value is obtained subtracting the risked value of the well *with* gravity and CSEM information (60 million dollars) minus the risked value *without* CSEM and gravity information (40 million dollars) minus the cost for acquiring, processing, and integrating CSEM and gravity data (5 million dollars). The result is 15 million dollars for one single exploration well.

It is intuitive that the value of an exploration well improves if the chance of success increases through integration of complementary geophysical data. In this simple exercise, I showed that the Bayes formula allows estimating the additional value obtained by this integrated approach.

6.3.3 The Generalized Value of Integration in Uncertain Domains

Handling uncertain knowledge in the decisional process represents the norm in many scientific fields. In these cases, integrating complementary information often represents the best approach for supporting our decisions. For instance, medical diagnosis is frequently affected by uncertainties; consequently, it is commonly based on combination of many types of information derived from independent sources.

Of course, physicians cannot use an infinite budget and infinite time for their diagnosis; thus, they cannot apply the whole spectrum of diagnostic techniques for collecting all the possible useful data. Moreover, their knowledge based on previous experience is commonly affected by uncertainties and it is necessarily incomplete. Consequently, they cannot pretend to apply an exhaustive and deterministic approach. It is more reasonable to apply probability principles and a Bayesian approach.

In many practical situations, the hypothesis of conditional independency helps scientists combining many different sets of independent observations in a simple, compact formula, for estimating the posterior conditional probability. We have seen that, in geosciences, the approach allows an easy formalization of a complex process: this is the combination of "real" observations (such as EM and gravity data) with our prior beliefs (such as geological models based only on seismic data). In medical diagnosis we can apply the same approach for combining different types of signs, symptoms, and images of body interior, together with our prior knowledge (such as the anamnesis, case histories, previous medical analyses, statistics, and so on).

It is important to remark that integration of information is a dynamic process: new information arrives into the "informative system" and it is continuously added to prior information. Thus, the practical problem is updating the status of knowledge through integration over the time. The so-called process of *Bayesian updating* is particularly useful in all those fields where there is the need of integrating a continuous stream of new data. This allows incorporating new evidences one piece at a time, modifying the previously held belief. As each new piece of evidence Y is observed, the belief in the unknown variable (for instance, a certain scenario Z) is multiplied by a factor that depends on the new evidence. In case of conditional independency between the set of new observations Y and the previous observations X, the *simplified Bayes rule for multiple evidence* is given by:

$$P(Z|X, Y) = \alpha P(Z) \cdot P(X|Z) \cdot P(Y|Z), \qquad (6.8)$$

where α is a normalization factor. This formula can be generalized to every number of conditionally independent observation sets. Interesting case studies can be found in Russell and Norvig (2003).

The Bayesian approach can be efficiently represented through a *belief network* or *Bayes network* (Heckerman et al. 1994). This is a probabilistic graphical model representing a set of random variables and their conditional dependencies. These variables are represented by nodes in the network; they can be observable quantities, unknowns, scenarios, or hypotheses. A set of arrows links couples of nodes. An arrow going from node X to node Y indicates that X is *parent* of Y.

Each node X_i is associated with a distribution of conditioned probability $P(X_i|Parent(X_i))$ that quantifies the effects of the parent(s) on that node. The graph has no oriented cycles. It is a type of *directed acyclic graph.*

Fig. 6.4 shows an extremely simple case of Bayesian network. "Atmospheric conditions" is independent from the other variables. Instead, "Symptom 1" and "Symptom 2" are conditionally independent but are caused by the same parent, "Disease X."

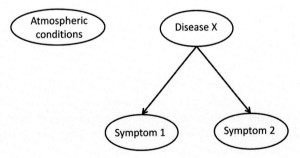

FIGURE 6.4   A trivial example of Bayesian network.

Every time we acquire new evidences, such as the observation of a new symptom, new blood analysis, new images of body interior, and so forth, the network becomes more complex. However, also our chances to perform a correct diagnosis theoretically increase. A successful diagnosis will be obtained if we will be able to integrate the new evidences properly. The Bayesian network together with the *simplified Bayes rule for multiple evidence* help us to represent the probabilistic causal links between the possible disease and many sets of complementary observations.

The beauty of this approach is its general applicability in many different contexts. For instance, it can be used for representing the conditional independency between two (or more) different types of geophysical anomalies (such as seismic, gravity, and EM responses), caused by the same parent, such as a hydrocarbon reservoir. When the conditional probability is assigned to each node, this approach provides a complete representation for the joint probability distribution in the domain under study (such as in hydrocarbon exploration).

Bayesian networks are used in many other fields, such as for modeling beliefs in computational biology and bioinformatics, document classification, semantic search, information retrieval, image processing, and in many decision support systems.[13]

6.4 SUMMARY AND FINAL REMARKS

Our brain is structurally, topologically, and functionally organized for integrating information. It continuously combines streams of heterogeneous data with previous beliefs, models, theories, feelings, and emotions. This attitude for integration represents a significant adaptive advantage. Combining complementary information can produce significant practical benefits, such as improving the decisional process in complex scenarios with many variables affected by uncertainties. The process of integration can happen at many different levels, ranging from basic unconscious multisensory perception to big data integration in complex scientific domains.

Of course, the process of integration plays a crucial role in exploration geosciences. In this field, integration of complementary data sets is often the only way for creating reliable models of the subsoil. Nowadays there is an extended scientific literature that demonstrates the benefits of

---

[13] Bayesian networks have been used for creating expert systems. For instance, Pathfinder is one of the first diagnostic expert systems used in medical sciences (Heckerman, 1991) based on a statistical approach for reasoning in uncertain domains. It was created for studying lymph node diseases by members of the Stanford Medical Computer Science program during the 1980s.

integrated geophysical models. For instance, in hydrocarbon exploration, combining seismic, EM, and gravity data allows robust estimations of rock properties of commercial oil/gas reservoirs. Moreover, in complex geological settings, integration of different geophysical methods helps to reconstruct the geological model and the structural features at extremely variable scale.

On the other side, integration is a process that requires additional human, economic, and technological resources. Consequently, it is important to evaluate, case by case, the cost/benefit ratio of this process. Especially when we operate in domains affected by large uncertainties, the Bayesian approach provides an effective method for combining prior knowledge with various sets of complementary observations. For instance, the simplified Bayes rule based on the hypothesis of conditional independency, allows estimating the *value of information* and the *value of integration* of many multidisciplinary data sets.

From a practical point of view, the process of integration adds effective value if it affects our decisions. Bayesian networks can effectively support the process of integration. They can drive the decisional workflow in complex scenarios affected by uncertainties, such as in medical, financial and business domains, and in exploration geosciences.

I would like to conclude this chapter with an important remark. Integration can happen in various forms not only by combining heterogeneous information (such as seismic and gravity data) but also by combining different sensory perceptions of the same source of information (such as images and sounds derived from the same physical signals). That type of integration can be called multiphysics—multisensory integration or *expanded integration*. This is possible; thanks to modern sonification techniques and cortical and subcortical structures of the brain dedicated to integration of multimodal information. The result is an expanded cognition of the data through audio—video display and dual-sense analysis.

The benefits of this multisensory interpretation approach are still largely unexplored. However, recent experimental results seem to encourage the development of new technology based on audio—video display. This interesting subject will be discussed in the next chapter dedicated to brain-based technologies.

References

Buland, A., Løseth, L.O., Becht, A., Roudot, M., Røsten, T., 2011. The value of CSEM in exploration. First Break 29, 69—76.
Dell'Aversana, P., June 2001. Integration of seismic, magnetotelluric and gravity data in a thrust belt interpretation. First Break.
Dell'Aversana, P., November 2003. Integration loop of 'global offset' seismic, continuous profiling magnetotelluric and gravity data. First Break 21, 32—41.

Dell'Aversana, P., Vivier, M., 2009. Expanding the frequency spectrum in Marine CSEM applications. Geophys. Prospect. 57 (4), 573–590.

Dell'Aversana, P., Carbonara, S., Vitale, S., Subhani, M.A., Otiocha, J., February 2011. Quantitative estimation of oil saturation from marine CSEM data: a case history. First Break 29.

Dell'Aversana, P., Colombo, S., Ciurlo, B., Leutscher, J., Seldal, J., 2012. CSEM data interpretation constrained by seismic and gravity data: an application in a complex geological setting. First Break 30 (11), 35–44.

Dell'Aversana, P., 2013. Cognition in Geosciences: the Feeding Loop Between Geodisciplines, Cognitive Sciences and Epistemology. EAGE Publications, Elsevier.

Dell'Aversana, P., 2014. Integrated Geophysical Models: Combining Rock Physics with Seismic, Electromagnetic and Gravity Data. EAGE Publications.

Dell'Aversana, P., Bernasconi, G., Chiappa, F., 2016. A global integration platform for optimizing cooperative modeling and simultaneous joint inversion of multi-domain geophysical data. AIMS Geosci. 2 (1), 1–31. http://dx.doi.org/10.3934/geosciences.2016.1.1.

Edelman, G.M., 1987. Neural Darwinism: The Theory of Neuronal Group Selection. Basic Books, New York, ISBN 0-19-286089-5.

Edelman, G.M., 1992. Bright Air, Brilliant Fire: On the Matter of the Mind. Basic Books, ISBN 0-465-00764-3. Reprint edition 1993.

Heckerman, D., 1991. Probabilistic Similarity Networks. MIT Press, Cambridge, Massachusetts.

Heckerman, D., Geiger, D., Chickering, M., 1994. Learning Bayesian Networks: The Combination of Knowledge and Statistical Data. Technical Report MSR-TR-94-09. Microsoft Research, Redmond, Washington.

Oizumi, M., Albantakis, L., Tononi, G., 2014. From the phenomenology to the mechanisms of consciousness: integrated information theory 3.0. PLoS Comput. Biol. 10 (5), e1003588.

Russell, S., Norvig, P., 2003. Artificial Intelligence: A Modern Approach. Pearson Education, Inc., publishing as Prentice Hall.

Tononi, G., Edelman, G.M., Sporns, O., December 1, 1998. Complexity and coherency: integrating information in the brain. Trends Cogn. Sci. 2 (12), 474–484.

Tononi, G., 2010. Information integration: its relevance to brain function and consciousness. Arch. Ital. Biol. 148, 299–322.

PART III

BRAIN-BASED-
TECHNOLOGIES
AND BRAIN
EMPOWERMENT

CHAPTER

7

Brain-Based Technologies

7.1 INTRODUCTION: FROM NEUROSCIENCES TO INNOVATIVE TECHNOLOGIES

In the first two parts of this book, I discussed the neurobiological fundamentals of several crucial aspects of exploration geosciences. Two main ideas motivated my analysis. First, I think that investigating the cognitive background of geodisciplines is interesting from a purely *humanistic* point of view. Before being geoscientists we are human beings; consequently, I assume that investigating the cognitive background of our activity is important. Second, *scientific* motivation is to use neurosciences for developing innovative technologies in geosciences. Indeed, neurosciences and biodisciplines have recently inspired innovative ideas in different scientific fields. Artificial neural network (ANN) and other "soft computing techniques" (SCTs) represent well-known examples of how cognitive sciences can drive the development of new information technologies (Aminzadeh and de Groot, 2006).

An interesting general review (not confined to geosciences) about "brain-inspired technologies" can be found in Hanazawa et al. (2010). The book includes contributions from several authors in the sectors of machine learning (ML), image analysis, pattern recognition, adaptive controller systems, robotic, artificial vision, chemical sensors, and so forth.

The IBM system called "Watson" is another famous example of information technology driven by cognitive concepts. This is a *question answering computer system* capable of answering questions posed in natural language (Fan et al., 2012; Finch et al., 2005; Kalyanpur and Murdock, 2015).[1] The original idea derived from an intriguing challenge: to build an artificial system that could compete at the human champion level in real time on the American TV quiz show, Jeopardy.[2]

Nowadays there is an increasing trend, in industry as well as in academic sectors, in developing "ML" approaches. ML is a sector of computer science that uses statistical (or mathematical) techniques to construct a model (or system) from observed data, rather than codifying a specific set of instructions that define the model for that data. In other

---

[1] Watson uses a novel paraphrasing algorithm based on an entity relation analysis of text. The approach is able to combine deep linguistic analysis and background resources (such as domain dictionaries) for detecting and matching entities and relations.

[2] "Jeopardy!" is a TV quiz show where three human competitors must answer rich natural language questions over a very broad domain of topics, with penalties for wrong answers.

words, ML algorithms iteratively "learn from data" for allowing computers to find hidden insights without being explicitly programmed to perform specific tasks. Banks and other business organizations in the financial industry, government agencies such as public safety organizations, universities, and industrial research centers use ML for many different purposes. Familiar examples of "applications" based on ML are the self-driving Google car, online recommendations based on customers' preferences extrapolated from social networks, fraud detection, and medical diagnosis. An increasing number of products and applications can be found in marketing and sales, health industry, transportations, and, of course, energy industry, including oil and gas exploration.

Neurosciences and biodisciplines can inspire, drive, or influence the development of computer technology in Earth disciplines too. In this book, I distinguish the two expressions "brain-inspired technology" and "brain-based technology" (BBT). In the second case, technology is developed *taking into account* human cognition, rather than being just *inspired* by the study of the brain. That technology consists of hardware, software, procedures, and workflows that are developed consistently with the human cognitive abilities and limitations. Other authors use expressions, such as *cognitive interpretation*, to group "cognitively consistent" techniques aimed at optimizing data analysis and interpretation. Many of these approaches are addressed to improve visual perception and cognition of geophysical data (Froner et al., 2012, 2013; Lynch, 2006; Paton and Henderson, 2015).

The same expression of "BBTs" is often referred to brain—computer interface (BCI), sometimes called a mind—machine interface, direct neural interface, or brain—machine interface. These technologies are commonly based on direct communication pathway between brain and some type of external device. Some among the main goals of BCIs are assisting, augmenting, or repairing human cognitive or sensorimotor functions (Nicolelis et al., 2000; Wolpaw and Wolpaw, 2012). I would like to remark that in this book, I use the same terminology of "BBT" for a different purpose. In the following discussion, BBTs are intended to be methods for data analysis, interpretation, and teaching based on (and driven by) cognitive criteria.

In this chapter, I start with an overview of well-known computation approaches inspired to the functioning of the brain and to biology. These include ANNs, fuzzy logic algorithms, and evolutionary computation algorithms. Furthermore, I discuss, briefly, some recent approaches aimed at optimizing data and model visualization, based on the functioning of human vision system. Then, I introduce a new approach for multimodal and multisensory analysis of geophysical signals. This approach is aimed at expanding the realm of geophysical data analysis taking in account for visual and audio perception. It combines the benefits of both imaging and sonification techniques with pattern recognition and automatic classification methods. Finally, I discuss the concept of quantitative integration system (QUIS). This is an integration platform allowing optimized workflows for combining complementary data. QUIS represents a typical example of BBT. In fact, it is developed consistently with the main integration functions of human brain, characterized by high connectivity of advanced processing modules.

7.2 SOFT COMPUTING TECHNIQUES: A BRIEF OVERVIEW

A set of methods emulating the human mind have been developed over the past few decades in geosciences for facing many computational problems not properly solved by conventional techniques. These new approaches are commonly grouped under the category called "soft computing techniques," in opposition to "hard computing" traditional approaches. A good review about SCTs, together with interesting real examples of application in geosciences, can be found in Aminzadeh and de Groot (2006).

SCTs include artificial intelligence methods such as ANNs, fuzzy Logic, evolutionary computing algorithms, perception-based logic, recognition technology, and hybrid methods. The main difference with respect to standard computational approaches is that SCTs are tolerant of imprecision, uncertainties, and ambiguities. Moreover, they can deal with problems where approximate reasoning and partial truth predominate. All these methods find ideal application in geosciences, medicine, and all those disciplines strongly based on interpretative processes in uncertain domains.

SCTs are both *inspired* by the study of the brain, such as ANNs, and *based* on human cognition, such as fuzzy logic algorithms. Furthermore, some SCTs are inspired by biology rather than by cognition, such as evolutionary computing (EC) algorithms.

ANNs mimic the behavior of biological neural networks forming the brain. In contrast with "conventional" mathematical algorithms well suited for linear programming, arithmetic and logic calculations, ANNs are effective to solve problems related to pattern recognition and matching, clustering and classification. The key concepts of this approach are derived from the original idea of John von Neuman (1958), who explained the behavior of neurons in terms of digital units: every neuron can be "on" or "off." In case of activation, it transmits signals to other neurons forming an aggregate. All the neurons show the same binary behavior so that the activity of the complete aggregate can be described by a sequence of logic binary functions.

Despite its extreme simplification, the neural model allowed the development of ANNs. In fact, at about the same time, the neurobiologist Frank Rosenblatt built a machine, the "Mark I Perceptron," which was able to detect and identify figures. The key idea of Rosenblatt was that the artificial connections between neurons could change based on a supervised learning process. Thus, the system is able to improve its learning process by changing the neuronal connections. The misfit between actual and expected output determines the change at every connection. This discrepancy is again propagated through the neural network and represents the necessary information for updating the weights of the connections.

In 1960, Bernard Widrow and Marcian Edward Hoff developed two neural network systems at the Stanford University: ADALINE (adaptive linear neuron) and MADALINE (multiple adaptive linear neuron). These produced the first practical results in the field of signal processing and noise suppression in telephone communications.

The evolution of ANNs continued over the following years with alternate successes and failures. A critical analysis of the basic concepts of ANNs is provided by Fodor and Pylyshyn (1988). In 1985, the American Institute of Physics started the Annual Conferences of Neural Networks for Computing. The conference of 1987, organized in San Diego by the Institute of Electrical and Electronic Engineers, was attended by more than 1800 participants. Nowadays, neural networks are applied in many different scientific fields. Furthermore, they help in understanding the cybernetic nature of complex intelligent systems, including the human mind as well as large human organizations.

The term "network" in ANN is referred to the interconnections between the artificial neurons belonging to the different layers forming the system. For instance, let us consider a system formed by three layers. The first layer includes input neurons. These send data via "synapses" (artificial interconnections) to the second layer of neurons; the third layer is formed by output neurons that receive the signals from the second layer. More layers of neurons can form systems that are more complex.[3] The synapses are regulated by "weights." These can remain fixed or can change over the time during the "learning process." However, what is a learning process in a machine? A computer program is said to learn from experience with respect to some class of tasks and a performance measure if such experience allows the program improving its performance on those tasks.

---

[3] "Deep learning" represents a recent evolution of ANNs. It is a set of techniques that allows parameterizing neural networks with many layers and parameters. These algorithms are able to train networks with multiple hidden layers (i.e., "deep nets").

An ANN can learn modifying the weights of the interconnections. In fact, an ANN is typically defined by three types of parameters: the interconnection pattern between the different layers of neurons; the learning process for updating the weights of the interconnections; and the activation function that converts a neuron's weighted input to its output activation. In general, the learning process is based on the minimization of a cost function, C. This can be a measure of how well the neural network performed to map "training examples" to correct output. The role of a "learning algorithm" is to explore the solution (model) space to define the network structure (connections and weights) corresponding to the smallest possible cost, C. The cost function will depend on the problem we want to solve. For applications where the solution is dependent on some type of experimental observations, such as in geophysics, the cost must necessarily be related to the data.

The learning process is not necessarily driven by a training data set. In fact, there are three main types of learning processes: *supervised, unsupervised*, and *reinforcement* learning. In the first case, the weights of the ANN will be inferred using a set of *labeled training data*. These data are used to construct or discover some type of predictive relationship. For instance, we can assume that a subset of our seismic data represent the seismic response of an oil-bearing sand formation. That assumption is commonly based on accurate calibration using borehole data. Then we use that known information for *teaching* the ANN to recognize similar features in the rest of the seismic data (far away from the wells).

Instead, unsupervised learning is aimed at drawing inferences from data sets consisting of input data without labeled responses. Consequently, the network searches for structures (correlations) within the data, without any external supervision. For instance, cluster analysis is used for exploratory data analysis to find patterns or grouping in data without using any type of training data set.

Finally, reinforcement learning is based on the concept of "reward." The feedback information to the network is a single bit (called *reinforcement signal*) indicating whether the output is right or not. This process allows the artificial network to determine automatically the ideal structure (connections and weights) within a specific context, to maximize its performance. In other words, the machine learns its behavior based on feedback from the environment.

Neural networks find many applications in modern geophysics. These include automated picking of seismic first arrivals, horizon tracking and automatic interpretation of seismic data, seismic classification, seismic data inversion, multiple suppression, principal component analysis, reservoir property estimation, reservoir monitoring, and so forth. A detailed description of all these applications can be found in Sandham and Leggett (2003). The fundamentals about computing with neural networks are discussed by Lippmann (1987).

Another set of SCTs is based on fuzzy logic. Different from classical logic that only permits binary choices, which are either true or false, fuzzy logic allows handling concepts such as partial truth in which the truth values of variables may be any real number between 0 and 1. For instance, in classical logic an element either belongs to a set or not. Instead, in fuzzy logic, an element belongs to a set to a certain "degree of membership." This fuzzy approach is appropriate for dealing with uncertainties, ambiguous concepts, and imprecise information in complex contexts. It applies well in all those circumstances where it is difficult or impossible to define sharp boundaries separating classes of objects, as it happens in many sectors of geosciences.

Moreover, fuzzy logic allows describing properties and features with linguistic qualifiers, such as "porosity is about 5%," or "saturation is approximately between 20% and 30%." Real examples and applications of fuzzy logic in geophysics are discussed by Aminzadeh and de Groot (2006) and Sandham and Leggett (2003).

While ANNs and fuzzy logic are, respectively, inspired by and based on the functioning of the brain, EC is another set of methodologies that try to mimic biological evolutionary criteria. Genetic algorithms are the most famous methods belonging to this category of soft computing approaches. Other methods are evolutionary programming, evolution strategies, classifier systems, and cellular automata. All of them simulate the processes of selection, mutation, and reproduction described by Darwin in his theory of evolution.

Genetic algorithms start with an initial population of candidate solutions for a given optimization problem. The goal is to make this starting population evolving toward better solutions. Each candidate solution has a set of properties called *chromosomes* or *genotypes*; these features can be mutated and altered. The evolutionary process is iterative and the population at each iteration is called a *generation*. The fitness of every individual in the population is evaluated for each generation, iteration after iteration; this fitness corresponds to the value of the objective function of that optimization problem. The individuals showing a good fitness are stochastically selected from the current population. Then the *genome* of each individual is modified through recombination and random mutation. The mutated individuals will form a new generation. The process is iterated and the algorithm stops, for instance, when a satisfactory fitness level has been reached.

Many applications of genetic algorithms and evolutionary programming have been performed over the past three decades in different geophysical fields, including seismic inversion, tomography, well log data analysis, and so on.

Each of the SCTs summarized above has specific benefits and intrinsic limitations. Neural networks are optimal in dealing with uncertainty and nonlinearity; they show good fault tolerance and high learning capability.

However, ANNs offer limited capability to use mathematical models and/or statistical information. Fuzzy logic is the ideal approach for managing linguistic expressions, handling uncertainties, and partial truth; but it shows low capability in learning and optimization processes. Genetic algorithms work very well for solving optimization problems characterized by complex objective functions, in difficult domains where uncertainties predominate. Moreover, they show good fault tolerance. Unfortunately, genetic algorithms show their weakness in linguistic manipulations, representing expert knowledge, and real-time operations. Thus, combining two or more techniques can optimize the advantages and reduce the disadvantages of each approach used individually. For instance, neuro-fuzzy methods are suitable for working in domains affected by large uncertainties, with imprecise and noisy data. That hybrid approach is ideal for combining machine computation efficiency with human attitude in managing ambiguous information. Another useful hybrid method is based on combination of ANNs with genetic methods. Sometimes, this approach is known as "structure-adaptive neural network." It guarantees strong learning ability and high performance in difficult optimization problems. Finally, fuzzy-genetic algorithms allow inverting ambiguous and imprecise data with high optimization capabilities. Additional combinations are possible, depending on the specific problems to solve, computation resources, and data uncertainties.

7.3 OPTIMIZING VISUAL COGNITION

In Chapter 4, I have explained that human cognition is strongly (not exclusively) based on mental images and maps. For that reason, accurate imaging and visualization represent crucial requirements for a good interpretation system. Not only in geosciences, but also in other fields, the quality of the output images depends on the quality of the input data set. Consequently, industry is moving toward more and more sophisticated acquisition techniques that allow recording the full 3D seismic (or electromagnetic) wave field in a complete range of offsets and azimuths.

Furthermore, full wave inversion and time/depth migration algorithms allow very accurate definition of the velocity fields and correct positioning of the seismic reflectors in the 3D space. In this workflow, *effective visualization* (in both data and model spaces) plays a role equally important as acquisition, processing, and migration techniques.[4]

---

[4] I intend "effective visualization" as that set of technologies that allow effective display, user interaction, comparison, superposition, combination, and manipulation of data and models, eventually belonging to different geophysical/ geological domains.

Effective visualization is related to the concept of *cognitive interpretation*, as well clarified by the words of Gaynor S. Paton and Jonathan Henderson, here reported integrally: "Interpretation of 3D seismic data involves the analysis and integration of many forms and derivatives of the original reflectivity data. This can lead to the generation of an overwhelming amount of data that can be difficult to use effectively when relying on conventional interpretation techniques. Our natural cognitive processes have evolved so that we can absorb and understand large amounts of complex data extremely quickly and effectively. However, these cognitive processes are heavily influenced by context and colour perception. Seismic interpretation can benefit greatly through better exploiting the positive aspects of visual cognition and through techniques designed to minimize the pitfalls inherent in the cognitive process. The interpretation of data also requires the ability to combine data analysis with knowledge and expertise that is held by the interpreter. It is this combination of visual perception techniques to see the information, combined with interpreter guidance to understand what is seen, that makes interpretation of seismic data effective..." (Paton and Henderson, 2015).

Powerful 3D visualization tools have been largely applied over the past two decades, replacing the traditional 2D seismic "wiggle picking." However, despite that unquestionable innovation, no significant efforts have been done for bringing together advanced interpretative technology and the power of human cognition. Interpreting huge seismic data sets is not just a matter of picking horizons with different colors. The main challenge is to move from the data space to the model space, transforming disorganized complexity (heterogeneous information) into organized complexity (coherent models) through a comprehensive interpretation workflow (Dell'Aversana, 2013). That process of transformation involves technology and high-level cognition. A frequent mistake is to focus all the efforts on the technological aspects of the problem, neglecting the cognitive background. The objective of *cognitive interpretation systems* (or brain-based interpretation systems) is to optimize the power of computation-based approaches taking into account human cognition (Lynch, 2006).

A crucial point is optimizing the visualization of data. Maximizing the information content of images supports the whole interpretation workflow, including anomaly detection, pattern recognition, and classification of geological features.

One way for optimizing human vision during interpretation is "explicit encoding." This is a technique that computes the relationships between objects and provides a visual representation of the relationship, not just the data themselves (Gleicher et al., 2011). *Color blending* is a type of explicit encoding by which three attributes are displayed simultaneously using a red—green—blue (RGB) color scheme (this can also be achieved with cyan—magenta—yellow and hue—saturation—value color schemes).

(A)   **(B)**

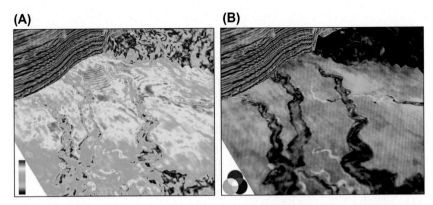

FIGURE 7.1   Comparison of two different representations of the same turbidite channel system. (A) Single attribute map and (B) red—green—blue blend of three frequency magnitude responses. *After Paton, G.S., Henderson, J., August 2015. Visualization, interpretation, and cognitive cybernetics. Interpretation 3 (3), SX41—SX48. http://dx.doi.org/10.1190/INT-2014-0283.*

This visualization approach allows us to visualize much more information than using a single attribute display. Color blending techniques are successfully applied in medical, astronomical, and geophysical domains.

Fig. 7.1 is an example of how an RGB color blend can reveal more detail than using one attribute alone. Panel (A) shows the map of a single attribute of a turbidite channel system, whereas panel (B) shows the same geological system using explicit encoding in an RGB blend of three frequency magnitude responses.[5] The second type of display allows an improved visual representation of the channels and their depositional geometries.

However, we can maximize the benefits of this visualization approach only if we are able to see and combine many different attributes easily and quickly. High interactivity in both data and model spaces is an additional key requirement for a successful interpretation system. In other words, simultaneous display of three or more geophysical attributes is just the first part of the job. The second crucial part is to combine and update images and models quickly. Consequently, a very important aspect of software design is the possibility "to easily combine different pieces of information to create the full picture" (Paton and Henderson, 2015). This can be obtained through "juxtaposition" (looking at different pieces of information, which are visually adjacent to each other), "superposition" (in which two images are overlain using opacity), and "explicit encoding" (where the relationship between different images is analyzed to produce a wholly new compound image).

---

[5] Interpretation is improved if the data are analyzed in both time and frequency domain. In fact, geophysical signals can be analyzed and interpreted within particular frequency ranges after proper frequency decomposition.

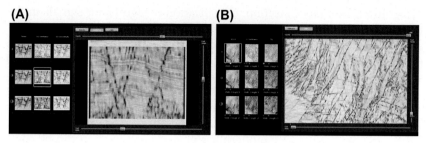

FIGURE 7.2   Examples of comparison of different attributes using juxtaposition, super-position, and explicit encoding. (A) Juxtaposition of different attributes and parameters with superposition (in section) of the attributes and reflectivity data. (B) Juxtaposition (in map) with explicit encoding of three attributes. *After Paton, G.S., Henderson, J., August 2015. Visualization, interpretation, and cognitive cybernetics. Interpretation 3 (3), SX41–SX48. http://dx.doi.org/10.1190/INT-2014-0283.*

Fig. 7.2 shows examples of (A) juxtaposition of different attributes and parameters with superposition (in section) of the attribute and reflectivity data and (B) juxtaposition (in map) with explicit encoding of three of those attributes (Paton and Henderson, 2015).

The high-level cognitive process of building a mental model of the geology is an iterative process. As we learn more the model evolves, as we test new scenarios the model evolves further. The ability of interpretation software to accommodate and to work with this cognitive process is a fundamental requirement of a *cognitive interpretation* workflow. To enable an interpreter to use their cognitive abilities effectively, interpretation software should:

- be interactive (by allowing variations to be visualized as the user is making the changes);
- maximize the information in the image (by allowing multivolume combinations such as color and opacity blends);
- enable options to be compared easily (using techniques such as juxtaposition);
- enable the interpreters to guide the process, to input their understanding of what is geologically feasible into the analysis and interpretation process.

Software that can provide all this functionality within a simple interface enables interpreters to use their high-level cognitive processes more effectively.

7.4 MULTIMODAL–MULTISENSORY ANALYSIS

Going back to the origins, the mammalian brain evolved over long time developing the ability to integrate and transform multisensory perception into high-level cognitive processes. These are, for instance, pattern recognition, anomaly detection, classification of significant signals in the frame of a wide context, and so forth. To survive, both preys and predators have developed the ability to detect, select, analyze, and interpret many signals in their environment. Consequently, it is reasonable to expect some benefits from *multisensory integrated cognition*.

The same concept is valid in science, especially in those disciplines where interpretation of heterogeneous information plays a crucial role. This happens in medicine, astrophysics, and, of course, Earth disciplines. All the benefits offered by modern systems of visualization can be complemented by advanced audio representation. Audio—video (or audio-visual) display of physical signals, including geophysical data, can be obtained through accurate analysis of data series (such as seismic traces, electromagnetic signals, well logs, and so on) in both time and frequency domain. This is what I call "multimodal—multisensory representation." In this section, I will summarize the fundamentals of this approach and I will discuss several real applications in exploration geosciences. Finally, I will discuss how multimodal—multisensory analysis can be implemented into comprehensive automatic systems addressed to big data mining, pattern recognition, and classification.

7.4.1 From Geophysics to Sounds

7.4.1.1 Introduction

The conversion of geophysical data into sounds can be done through many different approaches and at variable level of accuracy. If we desire to preserve the physical content of the data, advanced techniques of time—frequency analysis are required for avoiding artifacts. After transforming the data into the frequency domain, it is possible to use the derived spectrograms for creating musical files. These can be encoded in standard formats used in digital music, such as musical instrument digital interface (MIDI) or other common protocols. Finally, geophysical data sets can be played and interpreted using modern computer music tools, such as sequencers, digital synthesizers, digital mixers, and virtual musical studios (Dell'Aversana et al., 2016).

Auditory analysis of geophysical data is complementary and not substitutive of interpretation methods based on imaging. It usually involves the concept of sonification. This is a set of techniques used in several research fields to transform data into sounds and to represent, convey, and interpret them. Sonification has been used in geosciences since several decades (Benioff, 1953; Hayward, 1994; Kilb et al., 2012; Peng et al., 2012; Quintero, 2013; Saue, 2000; Speeth, 1961) and in other fields (see Dubus and Bresin, 2013 for a recent and exhaustive review).

The key requirement of a good transformation is to save the original information and, at the same time, to have a final file that does not occupy too much memory space. MIDI protocol is appropriate for assuring a proper time–frequency representation of a signal with limited memory requirements. On the other side, MIDI format cannot guarantee that the original information of the geophysical data is totally preserved. However, information loss is negligible if the time-to-frequency transformation is accurate.

7.4.1.2 Time–Frequency Transformation

Conventional Fourier transform (FT) does not guarantee any detailed transformation of nonstationary time series into the frequency domain. For instance, the frequency content of a seismic trace changes continuously with time, and the FT allows just a global representation of the frequency content of the whole trace. In other words, FT cannot highlight any local frequency contents and/or perturbations present in the signal. Consequently, it is generally unsuitable for studying the details of frequency variations versus time in nonstationary signals, such as the greatest part of the geophysical time series. If we desire to extract sounds from time series based on their "instantaneous" frequency content, we need to apply other types of time-to-frequency transforms. Among the various possibilities, two good candidates for that purpose are wavelet transform and Stockwell transform [or S-transform (ST)].

The wavelet analysis was introduced in the early 1980's for the analysis of seismic data. For a full list of references, see Foufoula-Georgiou and Kumar (1994). The key idea of the wavelet transformation is to use *windowing functions* that can allow only changes in time extension but not in shape. The integral wavelet transform is defined as

$$W_x(\tau, s) = \frac{1}{\sqrt{s}} \int_{-\infty}^{\infty} x(t) \cdot \psi^* \left(\frac{t - \tau}{s} \right) dt. \tag{7.1}$$

The wavelet transform is given by the inner product of $x(t)$ and translated-scaled versions of a single *windowing function*, $\psi(t) \in L^2(\Re)$. This function is called "wavelet." The coefficients $\tau, s \in \Re^+$ are called time delay and scale factor, respectively. Their role is to allow the wavelet changing its extension and amplitude as a function of the position of the windowing function along the time series. The "scalogram" is defined as the squared magnitude of the wavelet transform: $|W_x(\tau, s)|^2$. Different types of wavelets are applied in geophysics, such as Morlet, Gaussian, and Mexican hat (Shokrollahi et al., 2013).

S-transform is another type of transform that preserves the link with the Fourier framework and, at the same time, introduces additional

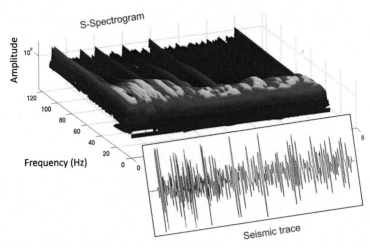

FIGURE 7.3   Example of a spectrogram of a seismic trace obtained through application of Stockwell transform. *After Dell'Aversana, P., Gabbriellini, G., Amendola, A., June 2016a. Sonification of geophysical data through time–frequency transforms, Geophys. Prospect.*

advantages. The original formulation of this transform applied to a generic time-dependent signal, *x(t)*, is (Stockwell et al., 1996)

$$S(\tau,f) = \frac{|f|}{\sqrt{2\pi}} \int_{-\infty}^{\infty} x(t) e^{-i2\pi ft} e^{-\frac{(t-\tau)^2 f^2}{2}} dt, \tag{7.2}$$

where τ is the time where the S-transform is calculated and f is the instantaneous frequency.

The S-transform offers several benefits with respect to other mathematical approaches. In fact, the windowing function scrolling the time series is not constant but depends of the frequency. Consequently, S-transform is appropriate when instantaneous frequency information must be preserved, such as in seismic signal analysis. Fig. 7.3 shows a visualization of a real seismic trace and the relative S-spectrogram $|S(\tau,f)|^2$, where $S(\tau,f)$ is calculated with Eq. (7.2).

7.4.1.3 From Time–Frequency to Musical Instrument Digital Interface Protocol

The next step is to transform the spectrograms derived from the time series into MIDI files. This is a standard hardware and software protocol to exchange information between different musical instruments or other devices such as sequencers. Typical MIDI attributes are the pitch, the velocity (sound intensity assigned to each MIDI note) and the note length; the timbre is instead not handled by the MIDI standard.

To convert a signal into a MIDI file, we have to do a sequence of operations to transform the physical quantities of a spectrogram (frequency, time, and amplitude) into the aforementioned MIDI attributes[6] (Dell'Aversana et al., 2016a). In summary, we convert the time–frequency into a time–pitch representation. The natural way to do this is to use the relation between the musical pitches and the frequencies covered by the musical instruments. Moreover, we make a correspondence between high sound intensity and high seismic amplitude. This approach allows very accurate transformation of geophysical signals, such as seismic traces and well logs, into MIDI files, with negligible loss of information.

7.4.1.4 Audio–Video Display

After creating MIDI files from our original geophysical data (such as a seismic trace or other types of data series), these can be automatically played using sequencer software. This is a sort of "interpreter" of the musical instructions codified in the MIDI file. The user can select one or more virtual instruments among many types of musical sounds included in digital synthesizers. In summary, the digital synthesizer can play each type of geophysical signal as MIDI instructions decoded by the sequencer. The execution speed can be set by the user, depending on the details to be highlighted in terms of musical patterns.

That audio representation does not exclude the conventional imaging techniques of analysis, but it is used as a complementary interpretation approach. Both images and associated MIDI sounds cooperate for improving our perception through a simultaneous visual–audio representation. Patterns of sounds organized in rhythmic, melodic, and harmonic structures emerge from the background, helping to detect anomalies and interesting signals, such as faults, stratigraphic features, hydrocarbon traps, and so on. Examples of this audio–video approach are discussed in detail in previous papers (Dell'Aversana, 2014b; Dell'Aversana et al., 2016a). I discuss again some of these examples to clarify the added interpretative value of this multimodal approach.

---

[6] The mathematical relationship between the frequency f and the MIDI note number n is the following:

$$f(n) = 440 \cdot 2^{(n-58)/12}.$$

In the equation above, the symbol n indicates the sequential number of MIDI notes. For instance,

n = 1 corresponds to the note C0 (16.35 Hz),
n = 2 corresponds to C0# (17.32 Hz),
n = 108 corresponds to B8 (7902 Hz)…

7.4.1.5 Application: Skrugard and Harvis Field

We applied our multimodal—multisensory approach to a seismic regional section (175 km long) acquired on the Skrugard—Harvis hydrocarbon discovery, in the Barents Sea, Norway. We used the wells drilled in the area to constrain our interpretation based on combination of images and sounds. First, we converted the seismic data from SEG-Y (standard format for seismic data) to MIDI format. Then, we uploaded the MIDI files into a software platform for digital music, including a sequencer and several digital synthesizers. Fig. 7.4 shows an example of multimodal display obtained by extracting the MIDI file at a fixed travel time. This audio—video display is available on YouTube at https://youtu.be/PeKy6qMv9Qg.

There is a circle moving from left to right along a linear direction fixed at 1200 ms. A few seconds after the start of the video, it is possible to hear the sounds associated with the image in correspondence of the position of the moving symbol.

The spectrogram (showed in the third panel from top) represents the frequency content of the amplitude envelop at 1200 ms (second panel). The user can listen to the MIDI sounds extracted from the seismic data in

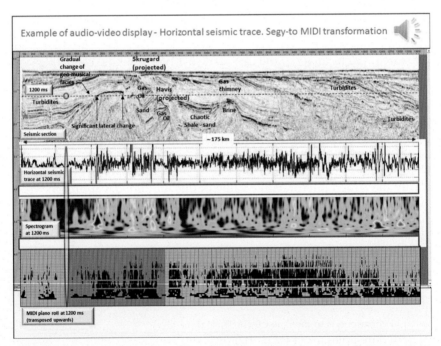

FIGURE 7.4   Example of audio—video display of a seismic section. *After Dell'Aversana, P., Gabbriellini, G., Amendola, A., June 2016a. Sonification of geophysical data through time-frequency transforms, Geophys. Prospect.*

correspondence of the position of the yellow circle. The user, using a sequencer, can select the MIDI instrument and the speed of execution. In this case, the execution speed of the MIDI file is very high. Despite the unavoidable dissonances, it is possible to listen to a great variety of musical patterns associated to significant seismic signals.

First, we can note that there is a good correspondence between sounds and images. Furthermore, there are many significant sound patterns also where seismic imaging shows low amplitude. This means that we can hear something that could escape from the view, like in the central part of the section just before the area of the gas chimneys. As expected, there are significant sound patterns in correspondence of the hydrocarbon reservoir. Thus, in principle, sounds can be used for reservoir detection (direct hydrocarbon indicator), after proper calibration. For instance, the pattern of sounds changes dramatically in correspondence of the "oil" label related with the Skrugard reservoir. Finally, there are significant sound patterns in correspondence of many other interesting geological features, such as gas chimneys, turbidite sequences, faults, and lateral geological variations.

In summary, sounds can support the interpretation of seismic data, when properly combined with "traditional" seismic images. Furthermore, different from any "conventional" frequency-based attribute, this auditory approach allows the perception of the whole frequency spectrum all at once for each time sample. The reason is that the MIDI file derived from the seismic spectrogram is a sequence of chords; each chord is an ensemble of simultaneous pitches and includes many frequency components. I remark that the same MIDI file can be played at much slower velocity, to capture much more details. An example of "slow MIDI execution" is shown at the following link: https://youtu.be/ld8CYt7eHug.

To verify the practical effectiveness of our approach, a group of geologists and geophysicists was asked to watch the audiovisual display of the Skrugard–Harvis section here discussed. The objective was to estimate (subjectively) the possible impact of this multimodal approach on the interpretation process performed by expert geoscientists. The main remarks of the experts can be summarized as following. First, there is a clear correspondence between imaging and musical attributes (100% of the experts agreed). Second, there are also many interesting musical patterns where seismic imaging shows low amplitude (100% of the experts). Third, the listening addressed the attention toward interesting signals that escaped from conventional visual interpretation, such as small faults (90% of the experts).

In conclusion, almost everybody in the test group agreed about the fact that the overall perception of the seismic information is improved when the imaging is analyzed in parallel with the MIDI sounds (90% of the experts).

7.4.2 Sound Pattern Recognition and Automatic Classification

The effectiveness of this audio—video analysis (alternatively named audiovisual display or multimodal analysis) increases if it is properly combined with automatic pattern recognition and classification approaches. We set a hybrid interpretation method based on both interactive audio—video display and automatic data classification. The workflow is divided into three main steps, as shown in Fig. 7.5.

First, we convert the geophysical data set into musical files (in both audio and MIDI formats), as explained earlier. Then we explore the whole data volume using automatic algorithms that are commonly used in the field of musical information retrieval (MIR). Finally, we perform audio—video display of selected subsets of our database, where special patterns of interest have been identified (as in the example of Skrugard—Harvis seismic data discussed earlier).

Our approach of automatic classification of seismic data using MIR algorithms is new in the geophysical domain, thus it requires some additional explanation.

7.4.2.1 Audio Versus Musical Instrument Digital Interface Features

The primary task in content-based MIR is to recognize/detect occurrences of a musical query pattern within a music database. In common

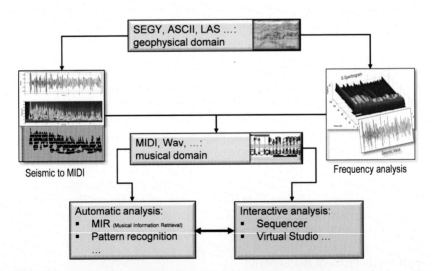

FIGURE 7.5   Hybrid approach for combining automatic classification and interactive audio—video display of geophysical data converted into images and sounds. *MIDI*, musical instrument digital interface; *MIR*, musical information retrieval. *After Dell'Aversana, P., Gabbriellini, G., Amendola, A., June 2016a. Sonification of geophysical data through time-frequency transforms, Geophys. Prospect.*

musical applications, the database consists of a collection of songs or other types of musical pieces. In our approach addressed to geophysical interpretation, the database consists of sounds obtained by conversion of geophysical data.

The work of automatic clustering and classification can be performed by extracting both audio and MIDI features (or attributes) from the data. In fact, musical data can be stored digitally as either audio data (e.g., Wav or MP3) or symbolic data (e.g., MIDI, GUIDO, MusicXML, or Humdrum). In the first case, analog waves are encoded into digital samples. Consequently, audio formats represent actual sound signals. Instead, symbolic data stores instructions and messages (for reproducing the sounds) rather than actual waves. Typical audio features are the pitch (that is closely related to the frequency) timbre, rhythm, energy (that is a measure of how much signal there is at any one time), and zero-crossing rate (that is a measure of how often the signal crosses zero per time unit). Other useful audio features are the cepstral coefficients. These are found from the FT to the log-magnitude Fourier spectrum and have been frequently used for speech recognition tasks.

Audio and symbolic formats each have their respective strengths and weaknesses. In the first case, the full waveform information of the data is preserved but at expenses of high memory requirement. On the other side, MIDI recordings contain just instructions to send to a synthesizer for reproducing the sounds. Consequently, the quality of the sound depends on the quality of the synthesizer. Moreover, as explained in the previous section, the original geophysical information can be preserved only if the transformation from geophysical signals to MIDI instructions is extremely accurate. However, MIDI format shows many advantages over audio recordings. They are much more compact, which in turn make MIDI instructions easier to store and much faster to process, edit, and analyze. Consequently, MIDI format is much more convenient than that of audio if we desire to extract precise musical information for classification purposes, such as sound patterns relating to notes, rhythms, and chords.

7.4.2.2 Examples of Piano Roll Display and Musical Instrument Digital Interface Features

MIDI format allows creating a discretized version of the spectrogram extracted from every type of data series. This can be visualized in the so-called "piano roll display," where the original signal is represented in terms of musical scores (see, for instance, Fig. 5.2). That type of representation allows searching for relationships between musical notes (vertical and horizontal relations, which means harmonic and melodic trends, respectively).

Finally, after accurate transformation from geophysical to MIDI format, many new features can be introduced. Some of these have no

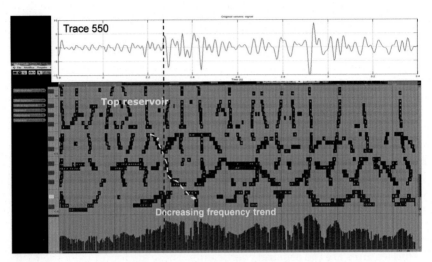

FIGURE 7.6   Example of musical instrument digital interface pattern in correspondence of a hydrocarbon-filled layer.

equivalent geophysical attributes. For instance, useful MIDI features are those based on musical texture, rhythm, density of musical notes (average number of notes per second), melodic/harmonic trends and note duration (such as average time between MIDI events, minimum, maximum, and average note duration), pitch statistics, and histograms of sound intensity.

Fig. 7.6 shows another example of MIDI display including interesting features. A single seismic trace is transformed into a MIDI file and is represented in the "MIDI piano roll display" (in the bottom panel, below the seismic trace). I recall that the vertical axis here represents the pitch (that is related to the frequency) using a virtual keyboard (showed on the left boundary of the figure). Different colors indicate variable sound intensity (red is high and blue is low). The horizontal axis represents the "MIDI execution time," which is proportional to the seismic travel time. The user can set the proportionality factor, regulating the execution rate of the MIDI file.

This particular seismic trace is interesting because it crosses a hydrocarbon reservoir. The correspondent MIDI display shows the histogram of the sound intensity versus time, in colors (at the bottom of the figure). In the piano roll display, it is possible to observe a clear sound pattern in correspondence to the top of the reservoir (marked by the vertical bar) showing a decreasing frequency trend. This is here interpreted as attenuation effect due to high hydrocarbon saturation in the sand layer forming the reservoir. Consequently, the MIDI trend can represent a useful feature for detecting hydrocarbon-bearing layers.

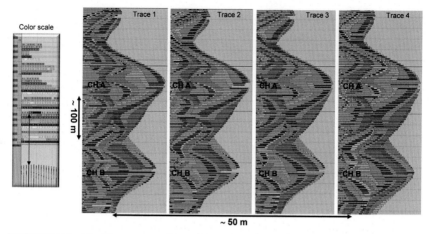

FIGURE 7.7   Pitch histograms of four adjacent seismic traces crossing two sand channels located at different depth. Vertical axis: depth (m); horizontal axis: distance (m). *Modified after Dell'Aversana, P., Gabbriellini, G., Marini, A.I., Amendola, A., 2016c. Application of Musical Information Retrieval (MIR) Techniques to Seismic Facies Classification. Examples in Hydrocarbon Exploration. AIMS Geosci. 2 (4), 413–425. http://dx.doi.org/10.3934/geosci.2016.4.413.*

Additional MIDI patterns can be observed in the same figure. This type of multivalued feature (formed by many musical scores) can be interpreted with the help of borehole data (available in this case). Features like this are diagnostic when used for training the process of ML in automatic facies classification. Of course, a single MIDI pattern can be misleading if it is used as stand-alone information for classification purposes. Thus, it must be used in combination with many other patterns and other MIDI features, possibly after proper calibration with well data.

Fig. 7.7 shows an example of four adjacent seismic traces, which are converted into MIDI format and displayed as pitch histograms. In this type of display, the vertical axis represents depth in meters. A segment of about 500 m has been selected for each trace, crossing two different sand channels (here labeled as "CH A" and "CH B"). Colors represent the different pitches (see the color scale on the left, related to the musical notes). The bar length in the histogram is proportional to sound intensity.

Pitch histogram is a useful MIDI "vectorial feature" (consisting of more than one MIDI score) that provides a discrete representation of the frequency content of seismic trace segments. It can be diagnostic for distinguishing different seismic facies. For instance, looking at Fig. 7.7, we can see that each individual channel shows its peculiar pitch histogram. Consequently, it is reasonable to expect that the pitch histograms can contribute to distinguish/classify different channels.

Fig. 7.8 shows another interesting example (extracted from a seismic data set recorded at a different location).

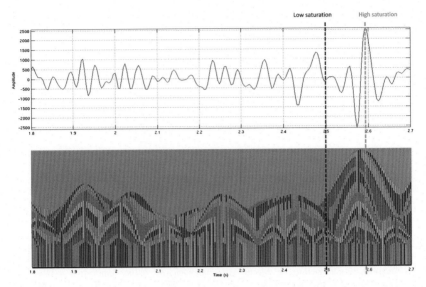

FIGURE 7.8   Seismic trace crossing a double gas-bearing layer with different saturation.

The upper panel shows a seismic trace extracted in correspondence of a double-layer gas discovery. The upper layer (highlighted by the red marker) corresponds to low gas saturation, whereas the lower layer (green marker) corresponds to high gas saturation.

The bottom panel shows the pitch histogram. The effect of high-frequency attenuation due to hydrocarbon is highlighted in the low-pass-filtered histogram of Fig. 7.9. Here, pink bars represent sound intensity of low-frequency pitches. This example suggests that filtered pitch histograms can represent helpful diagnostic MIDI attributes for distinguishing high-saturation from low-saturation gas-bearing layers.

More than 100 features can be extracted from the geophysical data converted into MIDI format. Of course, not all these MIDI attributes are necessarily useful for seismic facies classification. However, many different combinations of these attributes can be used for clustering the data in homogeneous classes.

There is not any absolute rule for selecting the optimal set of attributes to solve a specific classification problem. The choice of features that will work best depends on many factors, including the classification methodology being used, the type of data set, and the objective of the classification work. A common approach in terms of feature selection is to start out with a wide selection of candidate features and select progressively the ones that work well together for that specific problem. Another approach to feature selection is doing classifications of the training data set using different subsets of the features available and seeing which combinations perform best.

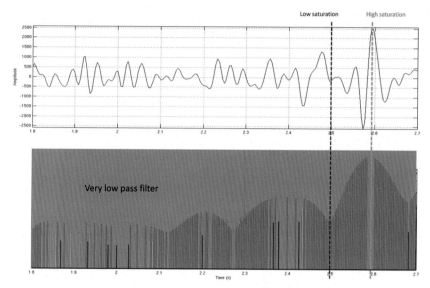

FIGURE 7.9   Low-pass-filtered pitch histogram. *Pink bars* represent sound intensity of low-frequency pitches.

7.4.2.3 Classification Steps: From Taxonomy to Training and Clustering

An important step of the classification workflow is to define a "model genre taxonomy." This corresponds to the desired number and type of categories in which the geophysical data will be grouped. Setting how many classes are necessary and sufficient for clustering a geophysical data set is a question affected necessarily by subjective decisions. The same happens when we desire to classify musical pieces forming a large database. In that case, we can start defining a musical taxonomy including "high-level" genres, such as blues, jazz, rock, pop, and so forth. After a first run of classification using these general categories, we can refine our workflow including musical subgenres (funky-jazz, for instance).

An analogous approach can be used for classifying geophysical data, after they have been converted into MIDI files. This type of format makes it easier to extract features relating to entire recordings because of the symbolic nature of the information encoded into MIDI messages. As I said earlier, the speed with which these messages can be processed represents a key advantage of MIDI versus audio formats. In the common practice, the classification workflow is improved iteratively using a trial and error approach. In fact, the classification results can be progressively improved using different classification algorithms.

A typical application is to cluster a seismic data set in different categories addressed to seismic facies identification (Amendola et al., 2017;

Dell'Aversana et al., 2016c). The data set is scanned trace by trace, with the objective to assign different data segments to distinct classes. We can start assuming that our data can be grouped into N distinct categories. Then, we can increase progressively the number of classes, refine our taxonomy, and, finally, improve the classification results.

We can decide to use supervised or unsupervised classification approaches; however, supervised methods seem to work better for this type of applications. In fact, training and testing our classification method on the data represent fundamental steps of the workflow. In the optimal case, we can define a training data set after proper calibration with borehole data. We can use a variety of classification methodologies, based on statistical pattern recognition and ML, including neural network and genetic algorithms.

Finally, a subset of our data (called "test data set") is used for evaluating the performance of the classification workflow.

We successfully applied this classification approach on real seismic data sets calibrated by wells drilled in correspondence of hydrocarbon discoveries (Dell'Aversana, 2014b; Dell'Aversana et al., 2016b; Amendola et al., 2017).

7.5 INTEGRATION SYSTEMS

In the previous sections, I described some technologies addressed to optimized imaging and multimodal analysis of geophysical data. These methods reflect some of the main high-level cognitive functions, such as imagery, imaging, multisensory cognition, anomaly detection, pattern recognition, and classification. Another key aspect of human brain is integration of information. Thus, a good BBT must reflect the crucial characteristic of our mind that is combining heterogeneous data into coherent models.

I have already remarked in different parts of this book that integration of multiphysics and multiscale data represents one of the main tasks in geophysics and in exploration geosciences. Many approaches and algorithms have been developed for combining seismic measurements, with other types of geophysical information, borehole data, and geological observations. The most recent trend is to combine several complementary methods into comprehensive software platforms that allow optimizing the integration workflow. I have discussed the general structure of these integration systems in a specific book dedicated to geophysical data integration (Dell'Aversana, 2014a). I introduced the concept of "QUIS." This represents a typical example of BBT. In fact, it is developed consistently with the main integration functions of human brain, characterized by high connectivity of advanced processing modules.

7.5.1 Embedding Modeling and Inversion Algorithms in Quantitative Integration System Platform

Owing to the intrinsic limitations of every individual integration approach, including joint inversion algorithms, QUIS combines different integration techniques (such as single-domain, constrained, cooperative, model-driven, and joint-inversion) in the same optimized platform. This system represents a sort of "second-order integration," i.e., an optimized combination of integration methods. Furthermore, QUIS allows performing effective data and model management, input–output flow, links between different algorithms, and, finally, strong interaction between people in collaborative environments.

Our brain combines massive information streams efficiently because its anatomy and physiology are organized for efficient multisensory imaging, real-time combination of mental maps, and quick data integration. The structure of a QUIS is based on the same criteria of efficiency, interconnectivity, and interactivity.[7] In fact, if the integration workflow was too time-consuming and if the cost-to-benefit ratio was too high, many geoscientists would prefer to avoid complex integrated approaches. That is true especially when they try to combine conflicting information for creating coherent Earth models honoring all the data.

Consequently, the first requirement of a QUIS is an effective visualization system that allows the users to combine, quickly and accurately, different types of geophysical/geological domains, in both data and model spaces.

Multidomain imaging plays a key role in data representation. The multiphysics database must be organized in such a way that the whole measurements are represented in the same multidimensional space, through corendered maps, multiple attribute cubes, and so forth.

At the same time, the users must be able to update Earth models and interpretation in relatively short time, when new data are progressively included into the workflow. Our cognitive activity is characterized by continuous feedback between basic perception and high-level cognition. Integrated interpretation of geophysical/geological data must be based on similar circular loops between data and model space. These loops are effective if they allow a quick update of old models with new data. This is possible if modeling and inversion are efficiently managed by multiple users that work in cooperation. Sharing their own expertise, they must be able to deal with multiple types of input and to combine different types of modeling/inversion processes in the same workflow.

---

[7] An ideal QUIS architecture should emulate, in some way, the same general organization of human cognition. Thus, QUIS is brain-inspired and brain-based technology at the same time.

The output of one process must be quickly addressed to trigger a new process, after proper format and/or mathematical transformation. For instance, the velocity field obtained through seismic tomography must be quickly transformed into a resistivity model for triggering electromagnetic modeling or inversion. Then, the output of the electromagnetic inversion must be used for improving the velocity model in the seismic domain. This approach is known as cooperative-sequential inversion loop. It effectively resembles the cybernetic feedback that characterizes many cognitive functions.

This type of circular approach can be performed in geosciences if a good library of rock physical relationships is available and if appropriate mathematical/format transformations from one domain to another can be applied. The circular process of format and parameter transformation must be simple and accurate at the same time. The users must be able to deal with multiparametric input/outputs flows without getting crazy with different types of formats, coordinates, and grids. Consequently, the concept of *shared Earth model* is crucial in a proper QUIS architecture. That model must represent the same target for different professionals (geophysicists, geologists, engineers, and so forth), who deal with multidisciplinary methods and data, at variable scale and resolution, combining different modeling/inversion approaches.

Another fundamental requirement for an effective QUIS architecture is that geoscientists must be able to combine, efficiently, new data with prior knowledge and uncertainties. This is possible if the users can import easily prior information in the model space. The update of the reference models to be used in modeling/inversion must be an easy and quick process. For instance, importing seismic interfaces derived from previous interpretation, well logs, geological constraints, and so forth is a fundamental requirement for proper model building. After creating a detailed starting model, Bayesian inversion allows introducing the complete prior information in the objective function.

Estimating the uncertainty distribution in the final models is another important need. A multiparametric geophysical model obtained through a complex integration workflow is not very useful if the uncertainty on the various parameters is not properly estimated. A good QUIS platform must include a module for calculating the sensitivity of the different data/methods of the workflow with respect to the different types of parameters. That sensitivity must be properly represented through maps, sections, and cubes, being a fundamental attribute for proper interpretation of the shared Earth model. Moreover, the uncertainty propagation from data to model space must be estimated. This can be done, for instance, using the Bayesian inversion approach that estimates the posterior distribution of model parameter covariance.

The whole process of integration has a statistical nature. In fact, all the main aspects of the workflow, such as parameter estimation by inversion, the link between petrophysical and geophysical parameters, uncertainty evaluation, and so forth, commonly show a statistical distribution. Thus, combining heterogeneous information in multiple geophysical/geological domains frequently requires significant application of geostatistical tools. These must be included in the QUIS platform in a specific geostatistical module that cooperates with the other parts of the same platform. That module plays a fundamental role in combining new data with prior information through cooperative forward modeling and inversion. In fact, all these processes are intrinsically statistical: they do not allow finding any deterministic solution, but a set of possible Earth models.

Finally, these multidisciplinary models must be interpreted. For instance, in case of hydrocarbon exploration, we commonly search for spatially consistent anomalies in velocity and resistivity models that can be associated with oil or gas reservoirs. Consequently, the module addressed to facies classification and pattern recognition represents another important part of QUIS platform. That module must be properly linked with the geological/borehole database, for allowing proper geological interpretation of the integrated patterns and anomalies.

Fig. 7.10 shows a scheme of the main modules of a QUIS platform. This simplified block diagram represents just the general structure and workflow of this type of comprehensive integration system.

The circular arrows included into the central block are aimed to highlight the circular nature of the processes happening in the system. In fact, the integration workflow is not linear: it commonly starts with a simple approach, such as data corendering and forward modeling. Then,

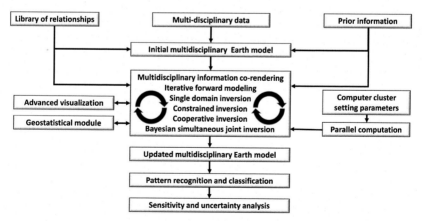

FIGURE 7.10　General scheme of quantitative integration system workflow.

the workflow becomes more complex, including sequential loops of single-domain, cooperative, and joint inversion. To have additional insight of the "cybernetic" loops between data and model spaces in a real-case history of geophysical data integration, I recommend reading a specific paper on this topic (Dell'Aversana et al., 2016b). This work includes theoretical details and real examples about Bayesian joint inversion of multidisciplinary geophysical data and QUIS platform.

7.5.2 Expanded Integration

QUIS platforms are currently applied in large-scale geophysical integrated projects (Colombo et al., 2014; Dell'Aversana, 2014a). This type of integrated system can be expanded including every type of geophysical, geochemical, and geological information. The basic QUIS architecture remains substantially unchanged.

Furthermore, the integrated interpretation workflow can be improved by multimodal–multisensory analysis, as explained in Section 7.4. This expanded integration approach can be represented by the same scheme shown in Fig. 7.10, adding the block diagram shown in Fig. 7.5. Looking at that QUIS scheme, we can see that all the typical formats commonly used in geophysics (SEGY, ASCII, LAS, etc.) can be transformed into musical files (MIDI, Wav, etc.). Consequently, multimodal–multisensory analysis can be applied to both data and models in the QUIS workflow.

For instance, we can extract sounds from single SEGY seismic traces, at the very beginning of the integration process. Alternatively, we can mine big seismic data sets transformed into MIDI sounds, applying MIR algorithms and automatic pattern recognition approaches during the phase of data interpretation. Finally, we can extract Wav or MIDI files at the end of the integrated workflow. We can apply multimodal–multisensory analysis to the final integrated models encoded into ASCII or SEGY formats, such as 3D cubes of seismic velocities, resistivity, or density models.

In summary, multimodal–multisensory analysis can be applied to both data and model space and to every type of geophysical measurement, parameter, or attribute. Combining multiphysical information with multisensory analysis allows expanding the concept of integration. This approach and its related technology are fully consistent with the intrinsic attitude of our brain to integrate heterogeneous information through multisensory cognition.

The following is just a short list of possible applications, some of which have been already successfully tested on real data (Dell'Aversana, 2014b; Dell'Aversana et al., 2016a, 2016c).

- Joint audio–video display and joint interpretation of composite well logs (sonic, resistivity, density, etc.).

- Frequency analysis of sonic and resistivity logs and joint inversion of the correspondent spectrograms for accurate estimation of porosity and saturations.
- Joint interpretation of audio–video display of multiple MIDI attributes extracted from seismic and nonseismic data.
- Big data mining, pattern recognition, and automatic classification of multiple MIDI attributes extracted by data belonging to multiple geophysical domains.

7.6 SUMMARY AND FINAL REMARKS

Earth disciplines, like almost all the other sciences, can progress by improving both technological and human factors. These aspects are strictly related. Indeed, developing new theories, algorithms, methodologies, and workflows is fundamental for solving the increasing challenges of exploration geosciences; however, the new technologies should take into account the functioning of human brain in cooperative environments.

Indeed, every new idea is never a magic event. It has deep biological, psychological, and social roots. It originates in our brain, never disjoined by our emotions, then it follows the path of complex cognitive processes, and, finally, it is shared within complex human communities and organizations. The primordial (mammalian) structure of our subcortical brain always works in background, strictly linked with our "modern" neocortex. This is a neurobiological fact with practical implications on the process of technological development. Earth sciences, like medical disciplines and other interpretative sciences, are strongly based on imagery and imaging, multisensory perception, pattern recognition, detection of weak signals, and classification and interpretation of integrated information. These are also key aspects of high-level human cognition. Part of the scientific community, in exploration geosciences as well as in other fields, has widely recognized the importance of these cognitive factors for the technological process.

Many innovative approaches emulating the brain have been developed over the past few decades, with successful applications. These are known as brain-inspired technologies. In this chapter, I have further expanded this concept, introducing new approaches that take into account integrated multisensory cognition. These ideas represent the background of new technologies here named BBTs. They are addressed to advanced imaging, interactive multimodal–multisensory data analysis, pattern recognition, automatic classification, and QUIS.

Despite the many analogies and similitudes, there is a substantial conceptual difference between brain-inspired technology and BBT. In the first case, technology is designed to substitute human mind in some type of computation tasks where machines can perform better than our brain. This happens, for instance, in big data mining; statistical analysis of large amount of experimental data; many cases of pattern recognition; and clustering and classification of images, sounds, and physical signals. Instead, BBTs are aimed at optimizing machine architecture considering the features of human brain and cognition.

This means that *BBTs are focused on human—machine cooperation rather than human—machine competition.* It could appear as a paradox, but excessive confidence in machine power could drive toward inadequate technological solutions. I am convinced that there is not any competition between the scientific/technological and the humanistic/philosophical cultures. We have just to optimize the way we create new technology without forgetting who the final user is: our brain.

In the remaining part of the book, I will show that the same key concepts concerning cognition and technology can support other important processes: education in geosciences, semantic approaches, and brain empowerment.

References

Amendola, A., Gabbriellini, G., Dell'Aversana, P., Marini, A.I., 2017. Seismic facies analysis trough musical attributes. Geophys. Prospect. http://dx.doi.org/10.1111/1365-2478.12504.

Aminzadeh, F., de Groot, P., 2006. Neural Networks and Other Soft Computing Techniques with Applications in the Oil Industry. EAGE Publications.

Benioff, H., 1953. Earthquakes around the world. In: Cook, E. (Ed.), On Out of This World side 2.

Colombo, D., McNeice, G., Raterman, N., Turkoglu, E., Sandoval-Curiel, E., 2014. Massive integration of 3D EM, gravity and seismic data for deepwater subsalt imaging in the Red sea. Exp. Abstr. SEG 2014.

Dell'Aversana, P., 2013. Cognition in Geosciences: The Feeding Loop Between Geo-disciplines, Cognitive Sciences and Epistemology. EAGE Publications, Elsevier.

Dell'Aversana, P., 2014a. Integrated Geophysical Models: Combining Rock Physics with Seismic, Electromagnetic and Gravity Data. EAGE Publications.

Dell'Aversana, P., 2014b. A bridge between geophysics and digital music. Applications to hydrocarbon exploration. First Break 32 (5), 51—56.

Dell'Aversana, P., Gabbriellini, G., Amendola, A., June 2016a. Sonification of geophysical data through time-frequency transforms. Geophys. Prospect.

Dell'Aversana, P., Bernasconi, G., Chiappa, F., 2016b. A global integration platform for optimizing cooperative modeling and simultaneous joint inversion of multi-domain geophysical data. AIMS Geosci. 2 (1), 1—31. http://dx.doi.org/10.3934/geosciences.2016.1.1.

Dell'Aversana, P., Gabbriellini, G., Marini, A.I., Amendola, A., 2016c. Application of musical information retrieval (MIR) techniques to seismic facies classification. Examples in hydrocarbon exploration. AIMS Geosci. 2 (4), 413—425. http://dx.doi.org/10.3934/geosci.2016.4.413.

Dubus, G., Bresin, R., 2013. A systematic review of mapping strategies for the sonification of physical quantities. PLoS One. http://dx.doi.org/10.1371/journal.pone.0082491.

Fan, J., Kalyanpur, A., Gondek, D., Ferrucci, D., 2012. Automatic knowledge extraction from documents. IBM J. Res. Dev. 56 (3.4).

Finch, A., Hwang, Y.-S., Sumita, E., 2005. Using machine translation evaluation techniques to determine sentence — level semantic equivalence. In: Proc. of the 3rd Int. Workshop on Paraphrasing. Jeju Island, Korea, pp. 17—24.

Fodor, J., Pylyshyn, Z., 1988. Connectionism and cognitive architecture: A critical analysis. Cognition 28, 3—71.

Foufoula-Georgiou, E., Kumar, P., 1994. Wavelet Analysis and Its Applications. Academic Press, Inc.

Froner, B., Purves, S.J., Lowell, J., Henderson, J., 2013. Perception of visual information: the role of color in seismic interpretation: First Break 31, 29–34. http://dx.doi.org/10.3997/1365-2397.2013010.

Froner, B., Purves, S.J., Lowell, J., Henderson, J., 2012. Perception of visual information: what are you interpreting from your seismic? First Break 31, 29–34.

Gleicher, M., Albers, D., Walker, R., Jusufi, I., Hansen, C.D., Roberts, J.C., 2011. Visual comparison for information visualization. Inf. Vis. 10, 289–309. http://dx.doi.org/10.1177/1473871611416549.

Hayward, C., 1994. Listening to the Earth sing. In: Kramer, G. (Ed.), Auditory Display: Sonification, Audification, and Auditory Interfaces. Addison-Wesley, Reading, MA, pp. 369–404.

Hanazawa, A., Miki, T., Horio, K. (Eds.), 2010. Brain-Inspired Technology, SCI 266. Springer-Verlag Berlin Heidelberg, ISBN 978-3-642-04025-2, pp. 29–32.

Kalyanpur, A., Murdock, J.W., 2015. Unsupervised entity-relation analysis in IBM Watson. In: Proceedings of the Third Annual Conference on Advances in Cognitive Systems (ACS 2015).

Kilb, D., Peng, Z., Simpson, D., Michael, A., Fisher, M., Rohrlick, D., March/April 2012. Listen, watch, learn: SeisSound video products. Electron. Seismol.

Lippmann, R.P., 1987. An introduction to computing with neural networks. IEEE ASSP Mag. 4 (2), 4–22.

Lynch, S., January 2006. Improving the interpretability of seismic data using achromatic seismic information. In: SEG Technical Program Expanded Abstracts 25 (1). http://dx.doi.org/10.1190/1.2369698.

Nicolelis, M.A.L., Wessberg, J., Stambaugh, C.R., Kralik, J.D., Beck, P.D., Laubach, M., Chapin, J.K., Kim, J., Biggs, S.J., et al., 2000. Real-time prediction of hand trajectory by ensembles of cortical neurons in primates. Nature 408 (6810), 361–365. http://dx.doi.org/10.1038/35042582.

Paton, G.S., Henderson, J., August 2015. Visualization, interpretation, and cognitive cybernetics. Interpretation 3 (3), SX41–SX48. http://dx.doi.org/10.1190/INT-2014-0283.

Peng, Z., Aiken, C., Kilb, D., Shelly, D.R., Bogdan, E., 2012. Listening to the 2011 Magnitude 9.0 Tohoku-Oki, Japan, Earthquake, Electronic Seismologist, N. of March/April. Cook Laboratories, Stamford, CT, 5012 (LP record audio recording).

Quintero, G., 2013. Sonificaton of Oil and Gas Wire Line Well Logs, International Conference on Auditory Display. ICAD.

Sandham, W., Leggett, M. (Eds.), 2003. Geophysical Applications of Artificial Neural Networks and Fuzzy Logic. Springer.

Saue, S., 2000. A model for interaction in exploratory sonification displays. In: Proceedings of the Sixth International Conference on Auditory Display. ICAD, Atlanta, GA, USA, pp. 105–110.

Shokrollahi, E., Zargar, G., Riahi, M.A., 2013. Using continuous wavelet transform and short time Fourier transform as spectral decomposition methods to detect of stratigraphic channel in one of the Iranian south-west oil fields. Int. J. Sci. Emerg. Technol. 5 (5).

Speeth, S.D., 1961. Seismometer sounds. J. Acoust. Soc. Am. 33, 909–916.

Stockwell, R.G., Mansinha, L., Lowe, R.P., 1996. Localization of the complex spectrum: the S transform. IEEE Trans. Signal Process. 44 (4).

von Neumann, J., 1958. The Computer and the Brain. Yale University Press, New Haven/London.

Wolpaw, J.R., Wolpaw, E.W., 2012. Brain-computer interfaces: something new under the sun. In: Wolpaw, J.R., Wolpaw, E.W. (Eds.), Brain-Computer Interfaces: Principles and Practice. Oxford University Press.

Web References

Audio-visual display of Skrugard-Harvis seismic section: https://youtu.be/PeKy6qMv9Qg.

Particular of the audio-visual display of Skrugard-Harvis seismic section: https://youtu.be/ld8CYt7eHug.

Applications to Education in Geosciences

8.1 INTRODUCTION

When I was a student in physics, my professor of mathematical analysis used to perform impeccable lessons writing long sequences of formulas on the blackboard, without adding any "useless" word or graphical explanation. He demonstrated complex theorems and solved difficult exercises using a rigorous and formal approach, without skipping any step. His teaching method was based uniquely on a rigorous mathematical formulation, avoiding deliberately any type of redundant verbal or pictorial clarification. The students of my classroom, including me, put their efforts in copying those formulas in their copybooks, at the same time trying to understand something. That was a hard work for young students, without the support of any additional "informal" explanation by the teacher.

Many years later, I suppose that the objective of my professor was to induce some type of attitude of autonomous thinking in our minds. Unfortunately, as confirmed by the final student tests after the mathematical course, that teaching approach was not very successful. In fact, most of the students failed the exam: we were not able to understand those formal concepts without the help of more intuitive representations, such as graphs and figures. In my case, for sure, the "challenging" approach of my professor did not work effectively.

Fortunately, I was able to understand his mathematical formulations many days later, only after linking those formulas with some type of physical concept. I remember that my understanding was possible, thanks to the help of my professor of physics. Different from the professor of mathematical analysis, he used a graphical approach for clarifying complex concepts of advanced mathematics and quantum physics. It is well known that theoretical physics includes difficult principles. These can be represented in multiple ways, such as using algebra formalisms, rules of symmetry, and graphical approaches. The American physicist Richard Feynman provided one of the best examples of how the mathematical formalism can be made intuitive by graphical representations. He introduced, in 1948, pictorial representations (the so-called Feynman's diagrams) of the mathematical expressions describing the behavior of subatomic particles. Feynman diagrams provide a simple visualization of the interaction of subatomic particles. This phenomenon would otherwise be described using a rather complicate mathematical formulation.

Feynman's diagrams are examples of how a complex matter such as quantum physics can be explained using a rigorous and intuitive approach at the same time. The basic cognitive principle of this "hybrid" didactic method is the same as that of the one used in brain-based technologies. Similar to the case of technology development, education and communication should be driven by criteria that are consistent with the functioning of human cognition. For sure, education should be supported, when possible, by graphical and/or other intuitive forms of representation.

We have seen in this book that the main high-level functions of our brain are based on imaging and imagery, pattern recognition, and multisensory integration. Consequently, also teaching complex concepts should be taken into account for those cognitive aspects. Scientific communication between professors and students, as well as between expert researchers and professionals, can be improved significantly if it is supported by images and sounds. Moreover, comprehension is extremely improved by recognition and integration. Especially novices learn better if they are able to recognize similitudes and affinities between new and old (consolidated) concepts. The same principle is valid for experts. Everybody is able to understand and store in his/her memory new concepts when these are properly represented, mapped, imaged, linked, and integrated with previous knowledge. Cognition is largely (not exclusively) a recognition process supported by a continuous mapping activity of the brain. Consequently, multisensory imagery, recognition of semantic structures, and integration of information should represent the fundamental bricks of education and teaching.[1]

All these concepts can be imported and applied in education in geosciences. In Chapters 5 and 7, I discussed the importance of auditory display techniques for representing data series as sounds. The well-known attitude of human brain in capturing patterns of sounds and integrating them into meaningful audio structures suggests that sonification, properly combined with imaging techniques, can represent an effective approach in education and training. In the following paragraphs, I will show how this method can be applied for explaining complex geophysical concepts in a simple way. Furthermore, I will discuss how audio–video display can be expanded in the frame of hypermedia technology.

8.2 APPLICATIONS TO SEISMOLOGY

Similar to theoretical physics, many geophysical concepts can appear difficult when explained through complex mathematical formulas. For instance, many students meet serious difficulties in understanding the physical nature of seismic or electromagnetic waves propagating through the Earth. Concepts such as spectrograms, frequency decomposition, Fourier analysis, high-frequency absorption, Earth filter, and so forth can appear quite obscure to students and beginners in geophysics. That is true especially if these topics are discussed directly (or uniquely) through equations. For instance, direct application of Fourier theory is not sufficient for explaining the frequency-dependent behavior of a nonstationary geophysical signal. On the other side, more advanced types of mathematical transforms

---

[1] The field of study related to learning improvement through brain-based approaches is also known as "brain-based learning." Following the definition of the glossary of Education Reform, "Brain Based Learning refers to teaching methods, lesson designs, and school programs that are based on the latest scientific research about how the brain learns, including such factors as cognitive development…" The Glossary of Education Reform, http://edglossary.org/brain-based-learning.

(Stockwell, Wavelet, etc.) can make the same matter even more complicate and difficult to understand.

An intuitive introduction using images combined with sounds can support the traditional rigorous didactic methods (Peng et al., 2012). Audio–video display of geophysical data such as seismic signals and well logs can be applied for these educational purposes. We have seen that seismic data, as well as any other type of data series, can be represented in terms of sounds derived from the frequency spectrum. I have already showed that the modern digital music software packages and the "virtual musical studio" technology offer many visual–audio tools for improving the comprehension of the physical nature of seismic signals. Students, young professionals, and senior geophysicists can take profit from this intuitive approach.

Fig. 5.2 shows an example of sonification of a single seismic trace. The top panel is the original seismic trace, whereas the central panel is the correspondent spectrogram. Placing side-by-side the same signal in both time and frequency domain is the first step for improving the comprehension of its physical nature. Furthermore, the bottom panel is the musical instrument digital interface (MIDI) file displayed in the "piano roll display" of a sequencer package. It shows the seismic trace in terms of combination of musical notes. This is a quite unusual and effective representation of a seismic signal. Indeed, it helps understanding the intrinsic spectral nature of the geophysical signal, showing the correspondences between its frequency spectrum and musical pitches.

This intuitive, but rigorous, representation of a seismic trace can be further improved looking at Fig. 8.1. The left bottom panel shows again the variable frequency content of the trace versus time. This time, differently from Fig. 5.2, colors represent the histogram of sound intensity associated with the different pitches (musical notes). In other words, this panel shows, through a pictorial approach, that every time series, such as a seismic trace, consists of many frequency components with variable intensity. For instance, we can select three different musical pitches on the virtual keyboard and we can see how the sound intensity changes versus time for each individual pitch (see the three panels on the right). This is a simple application of the method known as "spectral decomposition" that finds many applications in geophysics. Here it is illustrated using simple images and an intuitive link with familiar musical concepts (I provided other examples of pitch histograms and spectral decomposition of single seismic traces in Figs. 7.7–7.9).

This type of audio–video display represents a useful double-sense analysis of the frequency content of a nonstationary signal (such as a seismic trace). Many students and geoscientists confirmed that it effectively improves the intuitive comprehension of the correspondent physical phenomenon.

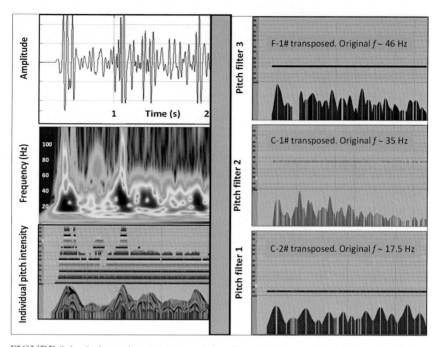

FIGURE 8.1   Left panels: seismic trace (top left); spectrogram (central left); colored histogram of pitch intensity (bottom left). In this last panel, every color represents the histogram of a specific musical pitch. Right panels: histograms of three selected pitches. These show the variations of sound intensity of individual musical notes versus time. The MIDI file has been transposed into the audible frequency range.

Furthermore, the advanced virtual studio technology available in digital music allows manipulating, editing, processing, and integrating the sounds, applying many types of user-friendly graphical tools. For instance, it is possible to combine a large number of MIDI files associated to multiple geophysical signals using a virtual mixer. This allows the user to see and listen to the sound of different types of data simultaneously. A typical application of this approach is "mixing" different types of well logs. Another useful approach is stacking the sound of MIDI files extracted from many seismic traces. Fig. 8.2 is an example of mixed MIDI seismic traces using a commercial sequencer package. Every seismic trace is encoded into MIDI file that can be played individually or together with other selected traces (Dell'Aversana, 2013). All the MIDI parameters (sound intensity, pitch transposition, musical instrument, special effects, and so forth) can be set by the user for each separate channel or for the whole ensembles of traces.

In summary, multimodal–multisensory analysis of geophysical data represents a powerful didactical approach. Modern platforms for digital

FIGURE 8.2   Example of virtual mixer combining several MIDI traces obtained by conversion of seismic data. *MIDI*, musical instrument digital interface.

music allow using intuitive pictorial/graphical representations of complex nonstationary signals, such as spectrograms, MIDI piano roll displays, colored pitch histograms, filtered histograms, virtual mixers, and so forth. These tools do not substitute the traditional mathematical formulation or software platforms for imaging analysis; however, they can support the education process in geophysics. This approach expands the perception of the geophysical signals and, finally, the comprehension of their physical nature.

There are many other fields of geosciences other than exploration geophysics where this didactical approach can be used. These include earthquake analysis and description of volcanic eruptions. In the next paragraphs, I will discuss an example in volcanology. I will show how the audio—video displays become more effective when they are supported by hypermedia technology.

8.3 MULTIMEDIA AND HYPERMEDIA IN GEOSCIENCES

Multimedia technology combines different types of contents including alphanumerical data, audio, images, animations, and video. Hypermedia (extension of the term hypertext) expands multimedia potentialities with interactivity and hyperlinks. The World Wide Web is the best-known type of hypermedia. Linking many different types of media improves the level of interactivity and increases the benefits offered by any individual technology. Many companies use approaches based on hypermedia for developing and/or promoting their products and services. A similar strategy can be implemented in geosciences, especially for educational purposes. The combination of different media and the web can be performed using quick response (QR) code technology. This allows expanding the possibilities of interactive and shared workflows in geosciences.

QR code is a technology developed in Japan by Denso Wave in 1996 (see Harish and Gurav, 2014 for a good summary of this technology). Initially it was created to allow high-speed component scanning to track parts in vehicles manufacturing. It is a type of two-dimensional bar code in the form of matrix code. It provides high capacity of encoding information. In fact, different from the one-dimensional bar code, it allows reading data in both vertical and horizontal directions. A standard type of QR code is a machine-readable optical label consisting of black modules (square dots) arranged in a square grid on a white background (Fig. 8.3).

It is used for fast information readability and immediate item identification, product tracking, marketing, document management, and many other applications. Nowadays it is used in many commercial fields addressed to mobile phone users. In fact, a smartphone or a tablet can be used as QR code scanners. The code is displayed and quickly decoded into textual information or URL linking to a website. Thus, the code represents an effective way to get useful information in real time about some type of target, such as a data set, a product, or a service. This is obtained through direct and immediate link with any type of media (images, movies, websites, blogs, animations, audio, and so on) available in the web. Any multimedia found on the Internet can be linked to a QR code. For instance, people can be directed to a specific YouTube or video clip using a QR link. The maximum data capacity is 7089 numeric characters and 4296 alphanumeric characters ("Version 40" of the QR code). There is significant research for expanding storing capacity of this type of code.[2] For instance, the High Capacity Colored 2D (HCC2D) code is able to increase storage capacity by using five different RGB colors.

FIGURE 8.3   An example of quick response code.

[2] I assume that this capacity will change by the time of publication of this book.

8.3.1 Example

8.3.1.1 Mt. St. Helens Eruption

In this paragraph, I would like to provide an example of how hypermedia technology can be used for improving the description of a complex geological phenomenon such as a volcanic eruption. Many different types of contents, files, formats, and media are linked together to combine the complementary value of heterogeneous information. This result is obtained using QR codes (Dell'Aversana, 2016).

An intense seismic activity and many phreatic explosions were recorded before the big eruption of St. Helens Volcano (State of Washington, United States) on May 18, 1980 (Brantley and Myers, 2000). The big eruption was preceded by a long series of steam-venting episodes and by volcanic swarms, caused by a continuous injection of magma below the volcano, at relatively shallow depth. Since March 1980, earthquakes and phreatic explosions were recorded, with increasing magnitude (3.2 or greater) during April and May. At 8:32:17 a.m., on May 18, a 5.1 magnitude earthquake triggered a huge landslide on the north face of the volcano. This exposed the magma in the volcano's neck to a sudden decrease of pressure. Consequently, the gas-charged rocks exploded causing an impressive lateral blast. In a few minutes, an eruption column of about 24 km expanded in the atmosphere. Pyroclastic flows (consisting of hot volcanic gases, ash, pumice, and pulverized rocks) started moving initially at 350 km/h and rapidly accelerated up to 1000 km/h. Volcanic lahars (mudslides that propagate a very high velocity) quickly reached the Columbia River, about 80 km to the southwest. Fifty-seven people and thousands of animals were killed and an area of hundreds of square miles was completely destroyed by the lahars and by the pyroclastic flows.

This eruption represents for sure a complex natural event involving many different scientific fields: geology, volcanology, geophysics, seismology, environmental sciences, human sciences, and so on. Let us suppose that we desire to introduce future geologists to the fascinating domain of volcanology using this eruption as a didactic case history. In fact, this eruption represents one of the best-documented volcanic activities. It can be described and analyzed with variable level of details using different types of media. For instance, the seismological precursors and the effects of the eruption can be studied through analysis of the seismic traces recorded by the surveillance geophone network deployed in the area around the volcano. In addition, a detailed analysis of the frequency spectrogram of the seismic activity can highlight additional physical features related with this eruptive event. Images and sounds of the eruption also can be very instructive, especially for students and researchers in volcanology.

A very interesting question is the following: is there any way to capture all that information in one synoptic representation? That question is crucial if we have educational objectives in mind. For instance, we could have the need to describe the big eruption in a textbook, or in a scientific paper, or in a poster to be showed at a scientific conference.

The answer to that question can be found using a combination of sonification techniques, QR codes and hypermedia. Fig. 8.4 shows an example of such type of hypermedia approach. It is a single figure including a huge amount of useful information encoded in many different types of media.

The top panel shows an interesting recording example (seismic trace in Wav format) of the intense explosive activity of the St. Helens eruption and before it.[3] However, just looking at the seismic trace is not sufficient

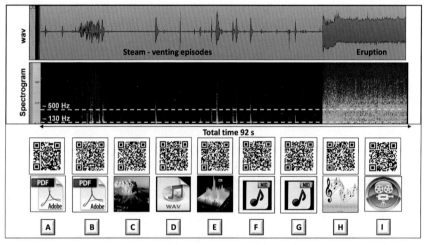

FIGURE 8.4   Example of hypermedia activated by quick response codes. This example is referred to the seismic activity associated with the volcanic eruption of Mt. St. Helens, 1980. A, description of the eruption; B, technical paper about geodata sonification; C, pictures of the eruption; D, "rumble" of precursors and eruption; E, 3D spectrogram; F, precursors played with MIDI instruments; G, eruption played with MIDI instruments; H, musical note sheet and "MIDI virtual piano" before and during the eruption; I, video of the eruption, by USGS. *URL and credits of videos about St. Helens Eruption of 1980: http://gallery.usgs.gov/videos/234#.VYvRDIKZd1Y.*

---

[3] Steve Malone of University of Washington, Seth Moran of United States Geological Survey, and Richard L. Hess of Audio Tape Restoration, Aurora, Ontario, have kindly provided these data. This file has been extracted from a data set recorded by seismometers deployed at different locations near the volcano. These recordings are sped up 128 times real time, so the frequency values are 128 times the actual.

for capturing this impressive natural phenomenon. Including sounds, images, descriptions, videos, and animations can be helpful for better understanding the sequence of the eruptive events. That information can be combined in the same figure. An immediate link between many different types of information describing the eruption can be obtained using different types of media files activated by QR codes. In this way, heterogeneous information can be quickly retrieved and easily combined.

The "sound" of the eruption for sure expands the perception of the sequence of events. I have included not only the Wav file of the original "rumble" of the eruption but also its precursors. I applied the "sonification" technique, described in Chapter 7 (Dell'Aversana, 2014; Dell'Aversana et al., 2016), to create MIDI files from the original Wav file. This can be done using the correspondent spectrogram, displayed in the second panel of the figure. After creating MIDI files, the audio can be analyzed using a sequencer. This is a type of software used for translating MIDI information into audible sounds. These can be analyzed at the desired execution velocity, to capture the details of every single sound associated with the eruptive phenomena.

Several clear spikes associated to seismic precursors can be observed in the spectrogram before the eruption. This starts in correspondence of a sharp variation in the frequency spectrum, with strong increase in signal amplitude in a wide range of frequency. The additional benefit of the MIDI file is that the sounds can be played automatically by music sequencer software at the desired velocity and using a selected digital musical instrument (music sequencers control virtual instruments implemented as software plug-ins, allowing musicians to replace real synthesizers with software equivalents). With this approach, every small MIDI sequence represents some physical event. Even though MIDI files can be edited using special software, they can be played by standard PC. Using both MIDI and Wav files, an additional perceptive dimension is added to the images and cognition is improved.

The Wav and the MIDI files associated with precursors and with the eruption can be activated using the correspondent QR codes, displayed in the bottom of the figure. (The MIDI files have been reconverted in Mp3 format. The correspondent QR codes specify the URLs pointing directly to these Mp3 files on the web.)

I labeled the QR codes associated with different types of information using sequential capital letters. The legend of the figure clarifies the content of each QR code associated with this eruptive event. Just scanning each QR code, for instance using a smartphone, the immediate decodification is performed and the multimedia information is captured.

I suggest to scan the QR codes (after zooming) in the figure using a smartphone or a tablet and to experiment directly the nice sensation to be projected into a hypermedia world. For instance, listening to the eruption played with MIDI jazz instruments at slow velocity (headphones are recommended) represents a unique experience. The MIDI sequence reproduces automatically the frequency content of the original Wav file, with transposition into the audible band. Melody, harmony, and rhythms reproduce the eruptive events in an inedited way. Many interesting musical patterns can be captured in the MIDI sounds associated with the eruption. Finally, I included in the QR code list the link to the US Geological Survey website, where documents, images, and videos have been published.

I am confident that this approach based on hyperlinks between different types of media can significantly improve the level of understanding of the eruption and its precursors. Finally, this is an advanced way to describe the eruption, with evident implications in teaching and education fields.

8.4 HYPERPOSTERS IN GEOSCIENCES

The example discussed in the previous section suggests an immediate application in the fields of education and scientific communication. For instance, the hypermedia approach here described can improve significantly the effectiveness of poster presentations.

Poster presentations represent an alternative medium to oral presentations for communicating scientific results during conferences and exhibitions. They have benefits and limitations. Among the benefits, we recognize that the relationship between speaker and audience is more direct and informal than that in oral communications. Furthermore, the poster is commonly left on the wall for a certain period (sometimes for the whole duration of the conference). Consequently, interested people can read the poster also before and/or after the time window scheduled for the presentation.

Among the main disadvantages, there is the fact that every poster must be necessarily synthetic. Too much information, figures, text, captions, tables, formulas, and so on could affect the message readability and its effectiveness. Everybody who attended a poster session had a sort of repulsion for those panels overpopulated by data, explanations, small-size text characters, complicate formulas, tables, and cryptic figures. Instead, poster showing just essential figures and synthetic text description appear for sure more attractive and can capture the attention of a larger audience.

The crucial question is how to balance properly the trade-off between the informative content of a poster presentation and its readability. Using QR codes and hyperlinks to multimedia represents for sure a promising approach for including many accessible scientific details into a readable poster (Dell'Aversana, 2016). For instance, a multimedia representation like that showed in Fig. 8.4 can be included into a poster with a very short caption. Everybody can access many details just using a smartphone and scanning the QR codes. This approach is facilitated by the static nature of the poster presentation. In such a way, an intrinsic limitation of this type of communication medium is converted into a benefit. This would not be possible during oral presentations. In fact, it is well known that mobile phones, cameras, and other devices like these are forbidden during oral sessions. In addition, the reference list can be encapsulated into a QR code printed at the end of the poster. In such a way, every reference can be linked directly with the web just through a simple QR scan, with the advantage of a real-time connection to the cited papers.

8.5 SUMMARY AND FINAL REMARKS

In this chapter, I discussed how combining sonification techniques and hypermedia can improve education and communication in geosciences. This approach improves the process of comprehension of complex phenomena in seismology, volcanology, and other Earth disciplines. It does not substitute traditional teaching methods based on formal and mathematical description, but it is complementary. The main benefit offered by this educational methodology is the intuitive nature of the pictorial representations combined with sounds related to physical phenomena such as wave propagation. Furthermore, QR codes allow fast link to the web between the different types of media. Using these codes, heterogeneous information can be combined in compact representations and shared on the web.

An application to the big eruption of Mt. St. Helens volcano happened in 1980 shows the potentialities of this approach for educational purposes in geosciences. Many other applications are possible with this method, such as improvements of poster effectiveness at scientific conferences. The same approach can be generalized to other scientific (and not scientific) sectors, such as training of students and professionals in medical sciences. In principle, every type of information can be combined into compact multimedia files and shared on the web in real time using QR codes. A future development of this approach is to import this technology into a fully immersive virtual reality, providing a multisensory expanded cognition of geological and geophysical phenomena.

References

Brantley, S., Myers, B., 2000. Mount St. Helens — from the 1980 Eruption to 2000. Report of USGS Numbered Series, Fact Sheet, 036-00. U.S. Geological Survey.

Dell'Aversana, P., 2013. Listening to geophysics: audio processing tools for geophysical data analysis and interpretation. Lead. Edge 32 (8), 980–987. http://dx.doi.org/10.1190/tle32080980.1.

Dell'Aversana, P., 2014. A bridge between geophysics and digital music. Applications to hydrocarbon exploration. First Break 32 (5), 51–56.

Dell'Aversana, P., 2016. Hyper-poster: A New Medium for Improving Communication in Geo-sciences, Expanded Abstracts of EAGE Conference and Exhibition, Vienna, 2016. http://dx.doi.org/10.3997/2214-4609.201601032.

Dell'Aversana, P., Gabbriellini, G., Amendola, A., June 2016. Sonification of geophysical data through time-frequency transforms. Geophys. Prospect. http://dx.doi.org/10.1111/1365-2478.12402 published on line.

Harish, N., Gurav, S.S., 2014. Embedding a large information in QR code using multiplexing technique. Taraksh J. Commun. 1 (1), 6.

Peng, Z., Aiken, C., Kilb, D., Shelly, D.R., Bogdan, E., March/April 2012. Listening to the 2011 Magnitude 9.0 Tohoku-Oki, Japan, Earthquake. Electronic Seismologist.

URL and credits of videos about St. Helens Eruption of 1980: http://gallery.usgs.gov/videos/234#.VYvRDlKZd1Y.

Video Credits

Video Producer: Stephen M. Wessells, U.S. Geological Survey.
1980 Eruption Footage: Don Swanson.
Original Graphics: Lisa Faust.
Interview Producer: Ed Klimasuskas.

Web References

The Glossary of Education Reform. http://edglossary.org/brain-based-learning.

From Information to Significance

9.1 INTRODUCTION

A brief summary of the discussion developed in the previous chapters can be useful for introducing a new important theme: transformation of fragmentary and heterogeneous information into coherent significances.

Our brain is a complex neurobiological system embedded in the body and strictly linked to the external environment. It works incessantly for combining images, recognizing patterns, discovering anomalies, and integrating information. We use experience and rational thinking for creating and updating knowledge about the "internal" and the "external" worlds. In parallel with our rational thought, our subcortical brain "feels" both the external world and the internal state of the organism (proprioception). Negative and/or positive emotions continuously emerge. These can be fear, panic, anxiety, rage, aggressiveness, sadness, happiness, enthusiasm, curiosity, and so on. Our emotional brain motivates, supports, and drives us to explore the environment and to develop models about it. The *primordial* neural seeking system, regulated by the dopaminergic pathways, pushes us to make new discoveries. These can be material resources, such as food, water, and energy, or mental resources, such as solutions, concepts, and ideas. During that incessant activity of exploration, we assemble data, observations, and experiences to formulate new theories about the reality inside and around us.

That complex process of building knowledge is a "semantic process." This is here intended as the cognitive and emotional path by which our mind infers significances from fragmentary experience combined with previous knowledge. Continuous physical interaction with the environment plays a fundamental role in the semantic process. When we are children, we learn and develop significances through our body. Then, we partially continue using the same physical approach for exploring the world for the entire lifetime. In general, empirical events and data are heterogeneous and fragmentary. Thus, we try to transform the disorganized complexity of empirical information into the organized complexity of coherent theories. On the other side, we use models and theories for interpreting new empirical data. This is the self-feeding loop between inductive and deductive reasoning.

That virtual semantic circle can happen at very different spatial, temporal, and conceptual scales. It ranges from the activity of finding practical solutions for simple problems to the activity of formulating new equations in complex scientific sectors. This *process of signification* represents the core of human mind, not only in science but also in the daily routine. For the human brain, it is natural to extract significances from experience through a process of generalization, at variable level of complexity. For instance, Newton inferred the gravitational law from measurements of the attraction force between different masses at variable distances. Moving to a completely different conceptual level, but using similar inference principles, a 1-year baby can infer nonverbal messages just by looking at the facial expressions of his/her mother.

On the other side, the same inference process so easy for a baby is a very challenging target even for a sophisticated computer. Why the semantic process is so natural for humans and so difficult to implement in a machine? The cognitive scientist Douglas Hofstadter stated that this question represents the so-called "barrier of the significance." He means that inferring semantic content from fragmentary and ambiguous information represents the real difference between human mind and machines.

For that reason, philosophers and scientists have been involved through the centuries in the dispute about "theories of meaning." These investigate the key questions related to the concept of *significance*. For example, what do we intend when we say to understand the meaning of a sentence? How do we assign semantic content to expressions of a language? Is the significance just a label that we "attach" to things, words, and events? Alternatively, is it the result of a much more complex process? What is the relationship between the process of signification and the process of induction? Does the meaning emerge as the result of an inference? What are the key neurobiological aspects of the process of signification?

The origin of the philosophical question of significance is rooted in the ancient works of Plato and Aristoteles. The modern theories of meaning started probably only with the work of the mathematician, logician, and philosopher Friedrich Ludwig Gottlob Frege (1848—1925). The investigation continued with the fundamental contribution of Ludwig Wittgenstein (1889—1951). The so-called "question of the significance" engaged many other philosophers in the 20th century, including Russell, Quine, Carnap, Chomsky, Putnam, Kuhn, and many scientists such as Heisenberg, Bohr, Einstein, and Wiener. Nowadays, this research involves the analysis of basic functions of the brain and the neurobiological aspects of human cognition.

From all these researches emerges that the question of the significance is related to the problem of how humans develop and share knowledge. Consequently, many sectors erroneously considered separate, such as semantic and epistemology, philosophy of language and science of communication, ontology, and fundamentals of mathematics, are reciprocally linked. The glue is the concept of significance. This plays a fundamental role in the Earth disciplines too.

9.2 SIGNIFICANCE IN GEOSCIENCES

I agree with Hofstadter about the importance of the concept of "significance." For that reason, I dedicated a book to this subject, with particular reference to my direct professional experience in geosciences (Dell'Aversana, 2013). The daily practice in geophysics and in geology shows clearly that the signification process is a complex combination of analytic and synthetic thinking. First, it is based on the *analysis* of intrinsic features of our object of observation. More details we know about a certain geological "object," deeper is the geological significance that we can assign to it. For instance, our understanding of a rock sample will improve with the details and the accuracy of our mineralogical, petrophysical, and paleontological analyses. However, if we want to understand the *deep* geological significance of our sample, the analysis of its intrinsic features is not sufficient at all. In fact, every robust inference is related to the comprehension of the *relationships* and the *role* of our sample within its geological *context*. Larger is that contextualization, deeper will be the geological significance that we can infer. We can say that we understand the geological meaning of the rock sample if we are able to interpret its mineralogical, petrological, and paleontological features with respect to the geological properties of the outcrop, the sedimentary unit, the geological region, and so forth. In summary, if we desire to understand the *deep significance* of "something," we need to understand its intrinsic features, together with the relationships between these properties and the context.

Consequently, there are many hierarchical levels of significance and of inference. These levels depend on the accuracy of the process of analysis and on the generality of the process of contextualization. The first, basic level is obtained by inferring relationships between information belonging to the same context (intradomain inference). For instance, if we desire to understand the meaning of some borehole measurements, we can try to infer a mathematical relationship by drawing cross-plots of composite well logs.

A higher level of significance (and of inference) is obtained by inferring some type of relationship between different domains. This is called *cross-inference*. It can be triggered by finding analogies and affinities between phenomena belonging to different domains. This type of high-level inference leads to an expansion of the significance range. It is a productive activity aimed at creating new knowledge and discovering new connections between different parts of the reality.

It can be obtained, for instance, by mapping different parameters in the same spatial domain. For instance, geologists and geophysicists are used to map different rock properties in the same 2D or 3D space, to find useful correlations, discover new relationships, and infer analogies between different geophysical domains. In my daily work, I frequently create multiphysical displays of different geophysical properties (elastic, electromagnetic, magnetic, etc.) in the same map or in the same 3D space. Using those corendered representations, I am able to find correlations between different geophysical responses. This integrated information can provide useful indications about the presence of an important target, such as a hydrocarbon reservoir.

This approach is very common in exploration geosciences, as well as in other interpretative disciplines such as medical sciences. When geoscientists explore the subsoil, they try to infer some type of coherent *significance* from multiple types of measurements and observations. That significance can be represented by a "geological model" or by any other type of "exploration target." Unfortunately, the process of inference is often complicated by the fact that the target of the research is located many kilometers below the surface, as it happens in hydrocarbon exploration. Furthermore, the observations are not necessarily homogeneous. They can be seismic data, electromagnetic and gravity fields, weak magnetic anomalies, complex geological outcrops, rock samples, and so forth. All these observations form a fragmentary/heterogeneous data set that must be integrated.

In previous works, I discussed in detail how geoscientists transform multiscale and multiphysical information into coherent Earth models (Dell'Aversana, 2014). This is much more than a mere technical question. In fact, in background there is the process of *inductive reasoning*: this can be defined as the combination of many evidences (for instance, experimental observations) for inferring some conclusion (such as a *model* of the subsoil). However, unlike deductive arguments, every inductive inference allows for the possibility that the conclusion is false, complicating significantly the work of exploration geoscientists.

We have to add that, in geosciences, the *data* are often affected by subjective interpretation. Think, for instance, to the interpreted horizons picked on a seismic section. The crucial point is that, similar to other interpretative disciplines, geosciences are generally based on *interpretation* and only rarely (or never) on absolute *data*. Unfortunately, not all geoscientists are aware about the difference between the two concepts of *interpretation* and *data*.

However, the intrinsic interpretative nature of the exploration activity is not necessarily a negative aspect: it can be used for investigating how our mind operates for extracting significances from heterogeneous information through inductive reasoning. Thus, geosciences can illuminate some of the key aspects of the semantic process. On the other side, disciplines such as semiology, philosophy of language, and epistemology can help geologists and geophysicists to understand their own work from a deeper point of view (Dell'Aversana, 2013). Finally, neurosciences help to investigate the neurobiological roots of the process of signification.

9.3 SIGNIFICANCE IN NEUROSCIENCES

Nowadays, quantitative measurements and accurate imaging of neural activity can support intriguing ideas that, just few years ago, were confined in the realm of work hypotheses. Over the past two decades, neurosciences contributed significantly to explain the mechanisms of formation of meanings in our brain, for instance, through accurate analyses of electrical neural activity.

Among the many published papers and books on this fascinating subject, I would like to mention the work of Walter Jackson Freeman III

(January 30, 1927–April 24, 2016), an American scientist with a multidisciplinary background.[1] He conducted pioneering research on how brains generate meaning, approaching the question from many complementary points of view, including neurosciences, biology, and epistemology (Freeman, 1992, 2000, 2003).

Freeman remarked that the information processing metaphor of the brain has dominated neurocognitive research for half a century. Unfortunately, it is not adequate for providing any reliable model of high-level cognition. Unlike computers, brains function hierarchically. That hierarchy includes the microscopic level of neural connections through electrochemical pulses of individual axons and mesoscopic and macroscopic levels of medium-to-large neural populations oscillating in synchrony. That fractal organization represents the neurobiological background for the functioning of the brain as a chaotic system in conditions far from equilibrium (Prigogine, 1980). That view is supported by recent advances in imaging technology. Indeed, Freeman remarks the experimental basis of his theory. He writes that "…The physiological evidence has been gathered from electroencephalograms (EEG) (Barlow, 1993) recorded with high density electrode arrays, intra-cranially on or in the brains of cats, rabbits, and neurosurgical patients, and from the scalps of normal volunteers. Signal identification and pattern classification have been done with high temporal resolution using wavelets (Freeman and Grajski, 1987) and the Hilbert transform (Freeman and Rogers, 2002). … Modeling with nonlinear ordinary differential equations indicates that the EEG patterns form by a 1st order phase transition (Abeles et al., 1995; Freeman, 1992, 2000; Tsuda, 2001; Freeman and Rogers, 2002), resembling condensation of a gas to a liquid …" (Freeman, 2003).

EEG and other imaging techniques allow neurobiologists measuring the electrochemical oscillations of energy that enable the brain to maintain its states far from equilibrium and at the edge of stability conditions (Freeman, 2003). Neurons operate in a collective modality through continuous variations in the trend of state variables. These are expressed in terms of pulse densities and dendritic current densities. From a neurodynamical point of view, significance emerges when these state variables converge toward chaotic attractors in a dynamic landscape of electrical potentials in response to external stimuli. These can be evidenced by EEG displays, such as in the examples showed in Fig. 4.1.

However, Freeman remarks "…no physical or chemical measurement of brain activity is a direct measure of meaning. Meaning can be

[1] I have already introduced some basic aspects of Freeman's work in Chapter 4. However, I would like to remark again his fundamental contribution in explaining, from a neurobiological point of view, how our brain creates significances.

experienced subjectively in oneself, and one can infer it in other from the behavioral context in which measurements are made, but one cannot measure it ..." (Freeman, 2003).

9.4 FROM SEMANTIC THEORIES TO SEMANTIC TECHNOLOGIES

Combining semantic theories and neurodynamic principles with epistemology can produce fruitful ideas in other scientific domains. In fact, one of the most promising sectors in computer sciences is represented by "semantic technologies." Massively based on machine learning approaches, these are aimed at extracting and interpolating meanings and significances from big data sets available in the Internet and/or in local databases. *Semantic web* is an example of this type of technology. It is "... a web of information derived from data through a semantic theory for interpreting the symbols. The semantic theory provides an account of 'meaning' in which the logical connection of terms establishes interoperability between systems" (Shadbolt et al., 2006).

Why do we need a semantic web? The crucial problem is that the ordinary web is not able to interpret and to link the information content for extracting significance. In fact, the current research engines work mainly using keywords, rather than semantic content of information. That approach prevents any type of intelligent utilization of web. Instead, an intelligent web should be able to process information automatically. The goal should be to infer some type of shared significance from data, emulating human capability to perform inductive inferences. That objective can be (at least partially) reached representing information in a more semantically structured way. In the context of the modern Internet, that means to extend the network of hyperlinked human-readable web pages by inserting machine-readable metadata and specifying the relationships between the pages. This allows automated agents exploring the web more intelligently, using not just information, but linked data.

The increasing importance of philosophy and neurosciences is fully justified in this emerging technology: defining the concepts of "meaning" and "significance," understanding the principles of inference, and how our brain works for extracting a semantic content from sparse information are all essential questions to solve before implementing an effective semantic technology in machines. Among the main goals, there are (1) improving computer efficiency for better exploring the web or big local databases; (2) inferring new knowledge from available data; and (3) enhancing man—machine interoperability.

In many recent works, the principal problem of creating a semantic web has been confined to define *ontologies*. These represent a set of knowledge-based, interrelated terms aimed at creating a sort of common dictionary of significance in a given domain. Their goal is to create the basis for using a common language between peers to avoid misunderstanding and improving intercommunication. Ontologies are often based using the logic of predicates by which sentences are expressed using triples structures formed by subject, predicate, and object. An established language for formalizing those triples is the Web Ontology Language and its underlying Resource Description Framework and Resource Description Framework Schema languages.

Creating ontologies represents just a part of the work for creating semantic technologies. For sure, it helps the work of classification and of communication inside a specific community. Using large and detailed ontologies allows everybody talking the same language. However, extracting significances from heterogeneous data is much more complex than just classifying information using the same dictionary. Contextualization and discovering relationships represent certainly other two crucial aspects of the semantic process.

The creation of a semantic web can be supported by tool for creating formal representations of knowledge, which help finding new relationships and understanding the context. For instance, *concept maps* (Novak and Musonda, 1991) are graphical tools for organizing knowledge in a structured fashion.

The idea of concept maps was developed for providing a model of the origin of the first concepts created during our childhood. Following Macnamara (1982), children develop the first concepts during the first 3 years from the birth. Recognizing regularities happening in the world around them, children develop the ability to infer general categories from particular observations. In parallel, they develop linguistic abilities for assigning symbolic "labels" to these regularities.

Concept maps include concepts, usually enclosed in circles or boxes of some type, and links between concepts, indicated by connecting arrows. The links are explicitly described by short texts near the arrows. More concepts can be linked forming a meaningful statement called propositions. These are also named *semantic units*.

Concept maps have a hierarchical structure: more general concepts include the more specific. This approach takes into account the functioning of our brain that works to organize knowledge in hierarchical frameworks. Thus, technologies and methods that facilitate representing knowledge hierarchically enhance the learning capability (Bransford et al., 1999; Tsien, 2007). In that sense, concept maps can be considered an additional example of brain-based technology because they reflect a key (hierarchical) aspect of our cognition.

Furthermore, cross-links are used for connecting concepts belonging to different semantic domains on the same map. Including descriptive examples, it is possible to help understanding difficult concepts in the map. Finally, concept maps can be used for preparing the background for including formalized concepts into a shared ontology.

9.5 SEMANTIC SYSTEMS AND SEMANTIC ENTROPY

Technology can support the process of knowledge development, but there is not any useful knowledge without information and people. Some scientists and philosophers predicted the possibility in a near future of an advanced form of artificial intelligence (AI) independent from human control (Kurzweil, 2006). However, at moment, the role of human beings in developing knowledge is still necessary (fortunately). Thus, the crucial questions do not concern just how to develop "intelligent" algorithms for "intelligent" machines. Additional important aspects are the psychological and social factors/implications of the semantic process.

An interesting open question is how individuals, small groups of researchers, technical teams, companies, and entire scientific communities combine new information streams with established systems of significances (concepts, models, and theories). That question becomes crucial when new information is in conflict with consolidated knowledge. This is a common situation in exploration geophysics in the oil and gas industry. Nowadays, integrating huge and heterogeneous data sets represents one of the main challenges in hydrocarbon industry.

For instance, new data sets can be in contradiction with the current geological models. This can generate conflicts inside the same team, in the entire organization, or even in a large scientific community, depending on the importance of the project. That conflict can trigger completely different behaviors in individual researchers and in large organizations.

To study that interesting phenomenon, I introduced the expression of *semantic system* (Dell'Aversana, 2013). This is intended to mean informative system comprising information plus human resources and technology, organized to extract coherent significances from heterogeneous information. Examples at different scales are working teams, research groups, scientific communities, oil companies, and the web community. Quantitative integration systems, where people work in cooperative environments for integrating geophysical data, are examples of semantic systems. However, there are many other examples outside the domain of geosciences.

Furthermore, I introduced the function called *semantic entropy* for monitoring the performance of the semantic system. This function depends on the percentage of information that the system is able to integrate into "coherent/structured semantic forms" (Dell'Aversana, 2013).[2] In geosciences, an example of "coherent/structured semantic form" is a geophysical model honoring the experimental measurements. Thus, in geosciences, semantic entropy can be used as a measure of the ability of the system to transform information into coherent significances represented by geophysical/geological models.

An Example

I have mapped the trend of semantic entropy versus time during real experimental projects, where multidisciplinary data sets have been progressively acquired in a time window ranging from few weeks to few months. Fig. 9.1 provides an example of trend of semantic entropy, here smoothed for making the graph more readable. I created this entropy graph during a real project of integration of multiple geophysical data sets, indicated in the Fig. 9.1 with the labels "A−D." In this case, the semantic system corresponds to the geophysical team acquiring and processing the data, plus the data itself, plus the technological resources used by the team.

We note that semantic entropy oscillates up and down. This trend depends on the ability of the semantic system to use its human and technological resources for integrating the heterogeneous information (old and new data). Let us summarize the history of this geophysical project using the function of semantic entropy.

During the first couple of weeks, before starting the acquisition of new data, entropy decreased. In fact, in this phase, the team worked for integrating all the available information, such as geological observations, previous maps, and bibliographic information. Geologists and geophysicists combined all the prior knowledge for building a preliminary Earth model of the investigated area.

---

[2] A simple definition of semantic entropy is $E_s = W \cdot \ln[d_{tot}(t)/d_{int}(t)]$, where $d_{tot}(t)$ is the total data collected at time t (in byte), $d_{int}(t)$ is the integrated data at time t (in byte), W is a weighting factor that takes into account the "robustness" of the integration. For instance, in an integrated geophysical project, the data can be seismic and electromagnetic measurements. We can think to integrate these multiphysics data sets by a joint inversion approach. The "robustness" of the integration is inversely proportional to the misfit between observed and predicted responses in the seismic and electromagnetic domains. The logarithmic function is used in the definition of semantic entropy for practical reasons. In fact, it allows limiting the dynamic range of the entropy function. This definition of semantic entropy is intuitive. It indicates just that information disorder can be decreased by clustering sparse data. For instance, a set of uncorrelated data is more disordered and has higher entropy than a set of well-organized data, properly integrated in relatively stable "semantic structures." Here I am using the term "semantic" because I assume that clustering information represents one of the factors in the process of formation of the significance.

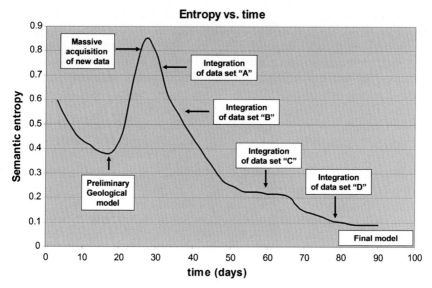

FIGURE 9.1　Trend of semantic entropy (smoothed) during a process of integration of multiphysics data. *After Dell'Aversana, P., 2013. Cognition in Geosciences: The Feeding Loop between Geo-disciplines, Cognitive Sciences and Epistemology. EAGE Publications, Elsevier.*

After 2 weeks, acquisition of new multiphysical data (seismic and electromagnetic) started. Entropy increased during acquisition simply because all the new experimental measurements were stored in the database as "field data," without any significant processing/interpretation. Consequently, the percentage of integrated information decreased with respect to unprocessed data. If we remember the definition of semantic entropy, it is clear that this function shows an increasing trend in this acquisition phase. We can see that there is a particular moment of "information overload" after about 4 weeks from the beginning of the project.

After almost 1 month from the beginning, the team started transforming progressively the field data into coherent geophysical models. Consequently, semantic entropy decreased. Noisy data were rejected and multiphysical measurements were progressively combined through modeling and inversion approaches. Thus, the percentage of integrated information increased over the time.

When all the data sets (seismic, electric, electromagnetic, etc.) were properly combined into a robust Earth model, semantic entropy reached its minimum, about 3 months after the beginning of the project.

I remark that this type of entropy trend represents just an estimation of the percentage of integrated information. In case of a geophysical project like the one discussed in this paragraph, the estimation can be performed measuring how many bytes of data are used for building geophysical/geological models. The robustness of the integration can be quantified using error functions, such as the misfit between observed and predicted responses. In other cases, when the amount of integrated information cannot be easily quantified, semantic entropy is much more difficult to estimate. Thus, it can be estimated only with large approximation. However, mapping semantic entropy over the time is generally useful. It allows monitoring the state of informative order in the system and its performance in the semantic process (Dell'Aversana, 2013).

An additional interesting result emerged from monitoring semantic entropy during real geophysical projects. It is remarkable that the peaks of this function of entropy correspond to a sort of "semantic crisis," where informative chaos dominates and people are unable to take decisions, or continuously change their models. Different human organizations can show completely different reactions to semantic crises. In fact, some organizations fail to restore the informative order and persist in a condition of strong instability: information remains in a disorganized form and the percentage of unstructured data increases over the time. This is a negative scenario for every type of organization. It can drive the semantic system toward a permanent chaotic situation, with all the negative implications that we can imagine.

In other cases, if the organization has sufficient resources for managing the semantic crisis, this can evolve into a positive situation. The crisis itself can trigger a *creative change of paradigm* at variable scales, depending on the context. In fact, sometimes, research teams, working groups, and larger human communities are able to use their resources in the best way exactly in correspondence of chaotic periods. They can use the instabilities created by conflicting information for changing the established significances.

Thus, a semantic crisis can represent the starting point of a real semantic *revolution* inside the organization itself. The result can be a radical change of the current models and theories, to face the informative chaos. In other words, if the old paradigm does not work, the system is constrained to change its approach, its models, and its systems of significances.

Thomas Kuhn highlighted very well that case of change of paradigm inside entire scientific communities involved in strong semantic crises (Kuhn, 1962). Semantic entropy is just a simple function of state by which these changes of paradigm can be monitored at variable temporal and conceptual scale. It highlights the dynamic relationships between information streams (data), formation of significances (models), and creative behavior (changes of paradigm) in semantic systems.

9.6 SUMMARY AND FINAL REMARKS

Extracting coherent significances from fragmentary experience is a crucial aspect of intelligence. The capability to do inductive inferences makes the difference between human mind and machines. It is true that some learning algorithms are able to perform an excellent work of pattern recognition, generalization, and automatic classification. However, I think that the maximum expression of the inferential process is still human creativity, where the semantic process drives individuals and/or entire human organizations toward new concepts, ideas, and theories.

The exploration workflow itself can be considered as a complex semantic process where creative inferences are essential for discovering new resources. Nowadays, industrial and academic communities are performing significant efforts for implementing semantic capabilities also in computers. This process is ongoing in exploration geosciences, as well as in medical disciplines and in many other scientific and business sectors. The semantic web is a well-known example of semantic technology.

However, despite the encouraging results, technology used independently from a correct epistemological approach is not sufficient for transforming complex information streams into meaningful knowledge. Although powerful algorithms and approaches have been recently developed for integrating heterogeneous data, the human factor remains crucial for moving from fragmentary information to coherent systems of significances. The main reason is that the human brain is still the most powerful "system" able to perform analogies and inferences. Indeed, the analogical and inferential thinking require much more attitudes than computation capabilities.

Differently from computer algorithms, human creative reasoning is based on both cognitive and emotive factors. These include expectations, esthetic sense, impulse toward new discoveries, and innate aspiration to model the reality. Furthermore, all these factors are driven by biological "adaptive values" necessary for our survival, including homeostatic regulation functions. Natural evolution forged these values in our complex organisms over many hundred thousand years. Our survival instinct is just the main evidence of that evolutionary process. Our pulse to explore the environment, discover new resources, build a refuge, escape from a danger, and play represents different forms of that basic instinct. Our intelligence is a unique mix of rational abilities and biological needs. It is made possible, thanks to the fact that the brain is embedded in a biological body and cooperates with other complex organic systems, such as the immune and endocrine systems. At moment, we cannot build any AI approach able to combine all these physical, cognitive, and emotional aspects into one inferential engine.

Consequently, my personal opinion is that technology *can support* cognitive, emotional, semantic, and creative functions of human mind but *cannot substitute them*. I introduced the concept of *semantic system* to reconcile the actual need to develop cognitive and semantic technologies with the role of human creativity in its social context. Indeed, the basic principles of semantic systems fit within the epistemological paradigm that has emerged mainly over the past century in science and philosophy of sciences: *information has small value without significance.* In complex scenarios, nothing can be identified as an absolute piece of reality independent from the rest of the system and from the human significances. These are based on social history, paradigms, assumptions, expectations, emotions, and intentionality. This concept of "significance" should be particularly clear to the majority of geoscientists who deal with sets of information that have no intrinsic value without human interpretation.

In conclusion, I think that the big challenge of the near future is to develop the right cultural approach for optimizing the interplay between human communities and technologies, rather than focusing all the efforts uniquely on computation power and artificial learning algorithms. The key message of this book is that geocognition can play a crucial role in this "scientific-humanistic challenge," through a new vision of the links between neurosciences and exploration geosciences.

References

Abeles, M., Bergman, H., Gat, I., Meilijson, I., 1995. Cortical activity flips among quasistationary states. Proc. Natl. Acad. Sci. USA. 92, 8616–8620.

Barlow, J.S., 1993. The Electroencephalogram: Its Patterns and Origins. MIT Press, Cambridge, MA.

Bransford, J., Brown, A.L., Cocking, R.R. (Eds.), 1999. How People Learn: Brain, Mind, Experience, and School. National Academy Press, Washington, DC.

Dell'Aversana, P., 2014. Integrated Geophysical Models: Combining Rock Physics with Seismic, Electromagnetic and Gravity Data. EAGE Publications.

Dell'Aversana, P., 2013. Cognition in Geosciences: The Feeding Loop between Geodisciplines, Cognitive Sciences and Epistemology. EAGE Publications, Elsevier.

Freeman, W.J., 1992. Tutorial in neurobiology: from single neurons to brain chaos. Int. J. Bifurc. Chaos 2, 451–482.

Freeman, W.J., 2000. Neurodynamics. An Exploration of Mesoscopic Brain Dynamics. Springer-Verlag, London, UK.

Freeman, W.J., 2003. A neurobiological theory of meaning in perception. Part 1. Information and meaning in non-convergent and nonlocal brain dynamics. Int. J. Bifurc. Chaos 13, 2493–2511.

Freeman, W.J., Grajski, K.A., 1987. Relation of olfactory EEG to behavior: factor analysis. Behav. Neurosci. 101, 766–777.

Freeman, W.J., Rogers, L.J., 2002. Fine temporal resolution of analytic phase reveals episodic synchronization by state transitions in gamma EEG. J. Neurophysiol. 87, 937–945.

Kuhn, T.S., 1962. The Structure of Scientific Revolutions. University of Chicago Press, Chicago, USA.

Kurzweil, R., 2006. The Singularity Is Near: When Humans Transcend Biology. Penguin Books.

Macnamara, J., 1982. Names for Things: A Study of Human Learning. MIT Press, Cambridge, MA.

Novak, J.D., Musonda, D., 1991. A twelve-year longitudinal study of science concept learning. Am. Educ. Res. J. 28 (1), 117—153.

Prigogine, I., 1980. From Being to Becoming: Time and Complexity in the Physical Sciences. W. H. Freeman, San Francisco.

Shadbolt, N., Berners-Lee, T., Hall, W., May 2006. The semantic web revisited. IEEE Intell. Syst. 21 (3), 96—101. http://dx.doi.org/10.1109/MIS.2006.62.

Tsien, J.Z., July 2007. The memory. Sci. Am. 52—59.

Tsuda, I., 2001. Toward an interpretation of dynamics neural activity in terms of chaotic dynamical systems. Behav. Brain Sci. 24, 793—847.

10

Neuroplasticity and Brain Empowerment in Exploration Geosciences

10.1 INTRODUCTION

Musicians and musicologists know the complexity of the opus of Johann Sebastian Bach. However, it is not necessary to be an expert in the musical domain for appreciating it. A mysterious form of beauty emerges in the compositional technique known as *fugue*. This polyphonic musical form was developed during the Baroque period and was chosen by Bach for expressing his genius in many unforgettable compositions.[1] It involves two or more voices in which a motive (subject) is exposed, in each voice, in an initial tonic/dominant relationship, and then it is developed by contrapuntal means.

Indeed, the Bach's fugues are extraordinary metaphors of the concept of *complexity*. This word derives from the ancient Latin term *complexus* that means "an aggregate of interconnected and related parts." A fugue expresses the same concept from a musical point of view. It generally consists of a series of *expositions* and *developments* with the purpose to reveal connections and relationships between seemingly unlike musical structures. The most important structural part of a fugue consists of the *main subject*. This expresses the primary melodic/rhythmic material of the fugue. The *head* of the subject is aimed at attracting the attention using particular rhythmic or intervallic emphasis. Then there is the *tail* of the subject, commonly more uniform from a rhythmic and melodic point of view. The fugue continues with a part called *answer*. This is a sort of imitation of the main subject, in a different voice and usually a fifth higher. Sometimes (not in every fugue), there is the *countersubject*. This commonly follows the subject or the answer and serves as *counterpoint*, sounding simultaneously in a different voice. Sometimes, *false subjects* are inserted in the fugue. These include just the start of the subject but are not completed.

Playing or listening to a Bach's fugue triggers deep musical emotions. However, it represents a fascinating cognitive experience too. Personally, when I play a fugue, I can feel my brain as if it is split in two parts but working as a whole. I need to use the right cerebral hemisphere for governing the movements of the left hand, and the left cerebral hemisphere for governing the right hand. Of course, all these movements must be perfectly coordinated in time and space for executing the main subject, the answer, the countersubject, and eventually the false subject. Consequently, both cerebral hemispheres must work with a certain degree of autonomy but in perfect synchrony. Moreover, my frontal lobes must be linked in some way to my temporal lobes for extracting a sort of "musical sense" from the sounds. In fact, not only I need to listen to the sounds but

---

[1] Of course, the fugue is not the only musical form used by Bach. For instance, there are many sublime preludes where Bach expressed his creative genius as a composer and his talent as a performer.

also I need to understand the musical patterns. I must catch the deep structure of the fugue, its subjects, and its answers, the countersubjects, and the false entries. Otherwise, I fail. In other words, I am not able to play the piece if I do not "feel" its architecture. My brains (I mean, left and right lobes) and my body must perform in harmony, following the rules of the "complexus" written on the musical arrangement. A good performance derives from the perfect combination between spontaneity and awareness, musical instinct and rational application of technique, synoptic view, and selective attention.

In those magic moments, I feel that my nervous system is fully interconnected and it works as a complex system. Furthermore, I feel the extraordinary sensation that new neural connections are created when, after many attempts, finally I am able to coordinate my hands, fingers, eyes, movements, thoughts, and emotions to realize a satisfactory musical performance.

All these factors create the well-known *power of music*. That power acts on our mind, changing it from many points of view. However, what do we mean when we say that a powerful experience, such as music, can change our brain? It is a well-established idea that the masterpieces of eternal composers such as Bach can have a strong impact on us. That is true not only from a cultural and emotional point of view, but also from a neurobiological point of view.

Indeed, our thoughts, experiences, and performances can change the structure and the function of our nervous system. Music is just one of the most familiar experiences through which we can modify the architecture of our neural connections. The good news is that synapses and neural maps can change significantly even into adult and old ages. This is the fundamental principle of the *neuroplasticity*.

10.2 THE BRAIN THAT CHANGES ITSELF

The idea of neuroplasticity emerged progressively starting from the end of the 19th century. It contrasts with the previous scientific paradigm about brain development. The "old" point of view was based on the dogma that the human brain remains relatively unchangeable after the critical period of impressive development during the early childhood. Neuroscientists thought that neural connections developed exclusively during the first few years of life. Consequently, if a particular area of the adult brain was damaged, the cognitive functions controlled by that area would be permanently lost because the neurons could neither form new connections nor regenerate.

A different opinion about the possibility of self-regeneration capabilities of the brain started emerging since the 1890s, when the psychologist

William James (1980) wrote in his book, *The Principles of Psychology*: "… organic matter, especially nervous tissue, seems endowed with a very extraordinary degree of plasticity …".

A few decades later, Karl Lashley (1923) performed experiments on adult rhesus monkeys by which he demonstrated changes in neuronal pathways after doing particular experiences. He interpreted these results as an effect of plasticity in brain circuitry. However, at that time, neuroscientists did not accept his hypothesis of neuroplasticity.

In 1949, Donald Olding Hebb argued that the neuronal connections are not static. They can be enhanced every time they are activated. This hypothesis is known as Hebb's rule. It implies that the learning process is not the result of a fixed property of the neurons; it is a time-dependent function of their variable connections. The basic idea at the base of Hebb's rule is that clusters of neurons tend to be excited together when a stimulus is perceived. Their oscillatory activity can continue after stopping the stimulus. In such a way, the event that produced the simultaneous oscillation of a group of neurons is fixed in the memory in terms of a selection of synchronized neurons.

Only in the 1960s, neuroscientists realized that stroke victims often regained certain cognitive functions if they performed adequate mental/ physical exercises after the ictus (under medical control). Bach-y-Rita showed that different areas of our brain could be activated for compensating different sensorial regions damaged by the ictus (Bach-y-Rita, 1967). This was a clear evidence that the brain is able to reorganize itself, modifying neural connections also in the adulthood. Other neuroscientists such as Michael Merzenich and Jon Kaas obtained significant experimental results supporting the hypothesis that the brain has a "plastic behavior" over the whole lifetime of a person (Merzenich et al., 1983a).

Nowadays the concept of neuroplasticity is largely accepted in neurosciences. A good historical review about this fascinating idea, together with interesting case histories about the capability of our brain to change its own structure, is provided by the psychiatrist and psychoanalyst Norman Doidge (2007). The author discusses how the neural circuits of the adult brain can be rewired during the entire lifetime, producing significant effects even more rapidly than previously expected. For instance, imaging analyses on the brain of university students preparing their exams have demonstrated that their gray matter increased in the posterior and lateral parietal cortex during a period of few months.

Indeed, the brain continues to create new neural pathways even in adult and old people. Existing synapses are continuously modified to adapt the function of the brain and the body to new situations and new experiences. This is the neurobiological mechanism by which we can recover damaged areas of the brain. This is also the way by which we can learn new information, acquire new attitudes, and develop new significances.

The reorganization of the synaptic networks of the brain happens in many different forms: not only as formation of new connections but also in terms of variability of the strength and efficiency of the existing neural connections. It is largely experience dependent, although it is influenced by genetic expression.

The crucial point is that neuroplasticity happens differently in every person depending on the individual's will to modify his/her habits and behaviors. Thus, the possibility of empowering our brain depends largely on our choices. We can decide whether to maintain our ideas and dogma unchanged forever or to open our mind and our life to new experiences. Consequently, the structure of our brain will change.

10.3 NEUROPLASTICITY: PRINCIPLES AND IMPLICATIONS

From a neurobiological point of view, brain plasticity is possible because the neural maps are dynamic. In other words, the topology of the brain changes continuously with experience, during the whole lifetime. Using dense microelectrode mapping techniques on primates, Merzenich highlighted how cortical sensory maps can be modified by experience. Together with Bill Jenkins and Gregg Recanzone, he demonstrated that sensory maps are labile into adulthood in animals performing operant sensory tasks (Merzenich et al., 1981, 1983b, 1984).

Experimental studies on neuroplasticity have many practical implications. If there is a causal link between our behavior and the structure of our brain, we can adopt active strategies for creating new connections between existing neurons and/or neural maps. Furthermore, we can create intentionally new maps and new systems of maps. In other words, we can define specific programs oriented to brain empowerment.

Many activities can promote new neural connections and/or reinforce the existing ones. I would like to introduce the fascinating hypothesis that the intensive and continuative practice of interpretative sciences, such as exploration geosciences, could induce significant effects of neuroplasticity. Indeed, there are significant lab evidences supporting that hypothesis. It is experimentally confirmed that the activity of exploration (in general, not specifically related to geosciences) can cause radical modifications of the brain structure of mammals (Panksepp and Biven, 2012). Following the same cause—effect relationship, geoscientists and other professionals involved in some type of exploratory activity could address their own neural modifications toward the right direction for empowering their brain. Multisensory imagery, pattern recognition, and semantic integration of heterogeneous information could represent crucial activities supporting neuroplastic changes. In particular, the daily application of brain-based technologies (BBTs) could support significantly that process. I will develop this intriguing hypothesis after explaining some basic neurobiological aspects of neuroplasticity.

10.3.1 Neurobiological Background of Neuroplasticity

The cognitive and neurobiological background of neuroplasticity stands in two basic concepts: *selective attention* and *integration* of different parts of the brain, respectively (Siegel, 2012). These two aspects are strictly linked and can cooperate for improving brain performances. In fact, we can activate selectively different regions of our brain depending on where we focus our attention. Furthermore, we can induce connections between these regions by performing activities that require integrated skills. A well-known example is provided by intensive musical practice. In fact, playing music can strengthen connections between the two hemispheres of the brain. Neurobiological research over the past few decades showed that professional musicians, who started playing before the age of 7 years, have developed a corpus callosum (see Fig. 2.2 as a reference) significantly thicker than normal (Elbert et al., 1995; Hyde et al., 2009). The corpus callosum is the bundle of nerve tissues that contains over 200 million axons; it allows efficient communication between the two cerebral hemispheres. It has been demonstrated that the structure of the brain can be significantly modified through constant exercise in a period of only 15 months of musical training in early childhood. The neural changes are well correlated with improvements in musically relevant motor and auditory skills. Furthermore, significant brain differences between children with different musical background have been found in frontal lobes, left-posterior pericingulate and left-middle occipital regions (Elbert et al., 1995).

These studies confirm that the activities that require selective attention and integration of skills can induce structural changes in neural connections. The neurobiological reason is that "neurons that fire together, wire together" (Shatz, 1992). This could be named *the fundamental principle of neuroplasticity*. Another basic principle is that "neurons work better (faster) when the size of the neural maps increases" (Jenkins et al., 1990; Recanzone et al., 1992; Doidge, 2007).

As I said earlier, there are important practical implications of these principles. We can work intentionally and systematically using selective attention and integrating information/skills to induce new neural connections in our brain and to improve synapses efficiency.

What are the biological mechanisms by which neural connections and neural maps can change, modifying the brain architecture? There are several possible processes: neurogenesis, synaptogenesis, myelino-genesis, and epigenetic effects.

Neurogenesis is defined as a process of generating functional neurons. This process is most active during the phase of development in the womb, but continues during adulthood in two regions of the brain: in the subventricular zone that forms the lining of the lateral ventricles (see Fig. 2.2 as a reference) and the subgranular zone that forms part of the dentate gyrus of the hippocampus area.

Synaptogenesis is the formation of synapses between neurons. It requires a neurotransmitter release site in the presynaptic neuron and a receptive field at the postsynaptic partners (see Fig. 2.3).

Myelinogenesis is the proliferation of myelin sheaths throughout the nervous system. It improves the connectivity between neurons and the connections between spatially separate brain regions that have sensory, cognitive, and motor functions.

Epigenetic effects are heritable changes in gene expression not involving changes to the underlying DNA sequence. They imply a change in phenotype without a change in genotype. Genic expression can be influenced by our experiences and consequent neural activation. Epigenetic mechanisms have been demonstrated through studies of neuro-development, learning, and memory. They have well-established roles in neuroplasticity within the normal brain. Moreover, epigenetic changes have been observed following stroke and during injury evolution (Arvidsson et al., 2002).

10.3.2 Basic Rules of Neuroplasticity

Neuroplasticity is governed by electrochemical phenomena between neurons. Consequently, it is subjected to physical and chemical con-straints. To have significant modifications in the topology of brain maps, several conditions must be verified. Based on recent research on the degenerative and regenerative effects of brain damage, Kleim and Jones (2008) summarized several *empirical rules* of experience-dependent neural plasticity:

- USE IT OR LOSE IT. Failure to drive specific brain functions can lead to functional degradation.
- USE IT AND IMPROVE IT. Training that drives brain functions can lead to enhancement of function.
- SPECIFICITY. Nature of the training experience dictates the nature of the plasticity.
- REPETITION MATTERS. Plasticity requires sufficient repetition.
- INTENSITY MATTERS. Plasticity requires sufficient training intensity.

- TIME MATTERS. Different forms of plasticity occur at different times during training and over the course of recovery.
- SALIENCE MATTERS. Training experience must be sufficiently salient to induce plasticity.
- AGE MATTERS. Plasticity occurs more readily in younger brains; adult brains are capable of plastic adaptation and some degree of structural organization.
- TRANSFERENCE. Plasticity in response to one training experience can enhance acquisition of similar behaviors.
- INTERFERENCE. Plasticity in response to one experience can interfere with the acquisition of other behaviors.

10.4 BRAIN CONNECTIVITY: MECHANISMS AND ENHANCEMENTS

During the second half of the 19th century, the work of the French anatomist and anthropologist Paul Broca, about the correlation between problems of aphasia and lesions in specific cerebral area, addressed neuroscientists toward a vision of the brain strongly based on *functional segregation*. From these studies, it appeared that the production of language is a cognitive function that is localized (in a right-handed person) in a specific area of the brain in the left lobe (Broca, 1861). That result, together with other evidences, suggested, at that time, that every specific cognitive function is localized with extreme precision in a dedicated portion of the brain. This paradigm about the functioning of the brain is known as *localizationism*. Functional segregation represented a dominant concept about the organization of the human brain in the 19th century.

Nowadays, a rigid localizationism is not compatible with a plastic vision of the neural system. The experiments of Merzenich, Bach-y-Rita, Elbert, and many other neuroscientists show that the different parts of the brain are interconnected. Furthermore, these connections are not fixed forever after the childhood. The modern techniques of brain imaging show clearly that the brain "talks with itself" through *dynamic* connections of the different cortical and subcortical parts.

10.4.1 Connection Mechanisms

Neural connectivity is the crucial aspect of the brain. It is a multiscale process that can involve individual neurons, entire neural populations, and large brain regions.

Connections of the various parts of the nervous system can arise following several mechanisms. Neurologists distinguish different types of connectivity. *Anatomical connectivity* is caused by structural connections such as synapses or fiber pathways. A different modality of neural integration is known as *functional connectivity*. It is related to statistical dependencies among remote neurophysiological events. This can be inferred by correlated activities of distinct brain areas (Gerstein and Perkel, 1969). Finally, the causal influence that one neural system exerts over another is known as *effective connectivity*. The distinction between functional and effective connectivity is not trivial. The functional connectivity can be quantified through measurements of statistical dependencies, such as cross-correlation and coherence. In contrast, effective connectivity is an intuitive notion of causal influence. It requires a model that tries to explain observed dependencies of different parts of the brain.

Sporns (2007) provides a summary of the methods for observing and estimating the different types of connectivity, including transcranial magnetic stimulation and functional neuroimaging. Skudlarskia et al. (2008) give another description of several imaging approaches, such as diffusion tensor imaging and resting state temporal correlations, both based on magnetic resonance imaging.

A fractal organization of functional brain networks emerges from functional connectivity studies in the frequency domain. Analyses of structural brain connectivity patterns show that our neocortex consists of clusters of densely and reciprocally coupled areas. Functional hubs have been identified, such as the hypothalamus. One of its most important functions is to link the nervous system to the endocrine system via the pituitary gland (hypophysis). Furthermore, the hypothalamus plays a fundamental role in sensing and integrating signals from the periphery and in maintaining homeostasis.

10.4.2 The Role of Neurotransmitters

Neural connectivity can be significantly modified by the action of specific neurotransmitters. I have already mentioned, in Chapter 3, that dopamine influences synaptic activity (Jai, 2003; Panksepp and Biven, 2012). It acts as a key neurotransmitter involved in different mechanisms of long-term potentiation (LTP) of neural connection. In fact, this neurotransmitter can influence the activity of ion channels and the functionality of membrane receptors affecting neural connections. Since 1973, studies on synaptic plasticity in the hippocampus and other brain areas have showed that the dopaminergic (DA) system plays a crucial role in synaptic plasticity and memory processes. Indeed, synapses that undergo neuroplasticity have been highlighted in various regions of the brain that receive innervations from the DA system. LTP effects have been observed in many regions in the neocortex and the striatum that receive DA projections from the ventral tegmental area, the adjacent substantia nigra, and/or retrorubral fields that contain the DA neurons (see Fig. 3.1 as a reference). In summary, dopamine transmission supports the process of associative learning, enhancing synaptic plasticity, and improving neural connections.

Rasmusson (2000) discussed the role of acetylcholine in cortical synaptic plasticity. This neurotransmitter can produce long-term increases in neural responsiveness. Furthermore, it can be crucial for restoring, at least partially, significant dysfunctions caused by loss or dysregulation of cholinergic inputs[2] (Kuo et al., 2007). Cortical cholinergic neuromodulations play an important role in the mediation of attentional processing, arousal, and other important cognitive processes. Dysfunctions can lead to cognitive impairments like those manifested in Alzheimer's disease. A discussion about hypotheses and experimental results about the attentional functions of cortical cholinergic neuromodulations can be found in Sarter et al. (2003).

An additional neurotransmitter that affects neuroplasticity is norepinephrine. It can influence the maintenance and reformation of neuronal networks. Moreover, the serotonergic system is thought to be a crucial mediator of neuroplasticity in depression. Serotonin [5-hydroxytryptamine (5-HT)] affects both morphology and neural activity of embryo. Furthermore, it influences neurogenesis and neuroplasticity after neuronal maturation, including proliferation, translocation, differentiation, and synapse formation (Veenstra-VanderWeele et al., 2000).

10.4.3 Neurotrophic Factors

Nerve growth factor is the first discovered member of the neurotrophin family in the early 1950s (Levi-Montalcini and Hamburger, 1951). This family of proteins promotes survival, growth, and function of neurons.

There are other proteins with trophic effects. The brain-derived neurotrophic factor (BDNF) was first isolated in the 1980s. It was shown to promote survival of a subpopulation of dorsal root ganglion cells, hippocampal and cortical neurons, peripheral sensory neurons, and nodose—petrosal ganglia (Barde et al., 1982). Since then, other members of the neurotrophin family such as neurotrophin-3 (NT-3) and neurotrophin-4/5 (NT-4/5) have been described. These show distinct trophic effects on subpopulations of neurons in the peripheral and central nervous systems.

BDNF plays a crucial role in molecular mechanisms of synaptic plasticity, supporting synaptic transmission/integration in the central

---

[2] Cholinergic neurons are specific nerve cells that send their messages mainly using the neurotransmitter acetylcholine.

nervous system, especially in the hippocampus and neocortex. It appears to strengthen excitatory (glutamatergic) synapses and weaken inhibitory (GABAergic) synapses. Copi et al. (2005) suggested that in vitro differentiated embryonic stem cells could behave as potential donor cells for cell replacement therapies of neurodegenerative diseases.

10.5 NEUROPLASTICITY AND EXPLORATION GEOSCIENCES

In Chapter 3, I assumed that the neurobiological background of exploration geosciences could be significantly regulated by the DA pathways. Furthermore, it is possible that other neurotransmitters involved in the play system, such as the opioids, acetylcholine, and serotonin, can influence the activity of geoscientists. In the previous paragraph, we have seen that both dopamine and serotonin are crucial for modulating neuroplasticity. Consequently, we can infer some type of connection, such as a possible circular causality, among exploration geosciences, specific neural systems, related neurotransmitters, and neuroplasticity.

Indeed, exploring the environment, as we do in geosciences, is based on the two key cognitive aspects of neuroplasticity: selective attention and integration. One could reply that the same aspects are crucial in many other scientific and artistic activities. That is true. I have already remarked that a good musical performance requires the capability to focus the attention and to integrate mind and body abilities. The same attitudes are required in other scientific domains, such as in medical sciences. It is realistic to think that all those activities that require some form of selective attention, interpretative attitudes, integration of skills, combination of fragmentary information, and intensive learning of new concepts induce some type of neural modifications. Without forgetting the other disciplines, I think that exploration geosciences imply a peculiar combination of activities, attitudes, and skills that, all together, can induce significant modifications in the brain architecture.

I have remarked many times in this book that geologists and geophysicists create, manipulate, and interpret maps, sections, and volumes of attributes. That continuous activity of imaging is related to a correspondent cognitive activity of imagery. Mental maps are combined and recombined in the mind of geoscientists during their daily work. The objective is to detect, select, and interpret specific signals that could be related to some type of target. Selective attention and integration of mapped information represent both the "daily bread" of exploration geoscientists.

Moreover, I assume that many technologies used in geosciences can support neuroplasticity. In particular some BBTs facilitate (and require at the same time) the cooperation of different cognitive skills. Consequently, the application of BBTs can produce new links and new neural connections between different areas of the brain. For instance, audio–visual display requires constant and focused attention using two perceptive senses, with simultaneous activation of visual and audio cortex on the same "object" (such as a seismic section).

Finally, the geological field activity, exploration, and outcrop analysis across an extended territory, correlation, extrapolation, and interpolation of observations represent a unique ensemble of factors that can stimulate neuroplasticity changes in the geoscientists' brain.

Let us analyze all these geoexploratory activities in some detail and their possible neuroplasticity effects.

10.5.1 Integrated Geophysics and Neuroplasticity

In Chapter 7, I have discussed the concept of quantitative integration system (QUIS). This is an example of BBT aimed at inferring multiparametric models from multiphysics data sets. While using a QUIS platform, people interact and cooperate for extracting coherent geological significances from fragmentary/heterogeneous information. This activity performed over a long period can have a positive impact on the brain. Linking information and cooperating with people is the best way to create new neural connections and to improve brain performances.

This concept has been widely discussed by the neurobiologist Daniel J. Siegel in his publications about "interpersonal neurobiology" (for an easy overview, see Siegel, 2012). The key concept of this interdisciplinary approach to the study of the mind is *integration* in all its aspects: combination of data, relationship between concepts, links between different domains of knowledge, cooperation between people, and, finally, connection between different parts of the brain. Siegel's message is that integration involves the linkage of separate aspects of knowledge and, at the same time, of different brain's parts. We can apply the same concept to the domain of geosciences: when we link different geophysical, geological, and geochemical models, at the same time we are linking different parts of our brain through new synaptic connections. What we do in terms of data integration reflects into neural integration.

The Siegel's approach includes also a methodology, called "mindsight," for improving the link between the emotional brain with the rational cognitive processes happening in the neocortex. "Mindsight" is the way we can focus our attention and awareness on our own thoughts and feelings. The objective is to potentiate the process of brain integration for improving our "internal world" and the relationships with the "external world." This could appear as an exotic approach; however, it has robust neurobiological fundamentals. Indeed, it is well known in neurosciences that our emotions originate continuously in the ancient mammalian brain that interacts with the neocortex. The modern sciences of neuroplasticity (including interpersonal neurobiology) teach that the interaction between subcortical and cortical brain can be improved intentionally empowering the brain as a whole.

In the activity of exploration, that neural interaction happens necessarily, although I doubt that geoscientists are aware of it. In fact, the DA system that regulates our seeking instinct originates in the ancient brain but innervate the neocortex (see, for instance, Fig. 3.1). Full awareness of that integrated process could represent a step forward for improving the exploratory skills and, finally, for empowering the brain.

10.5.2 Multimodal–Multisensory Analysis and Neuroplasticity

Multisensory integration allows us perceiving a world of coherent perceptual entities. It is an innate cognitive attitude that is crucial for the adaptive behavior. In Chapters 7 and 8, I discussed how that instinctual capability can be used in geosciences, through an approach based on multimodal–multisensory analysis of geophysical data. Now, I would like to expand the range of benefits potentially offered by this multimodal approach to data analysis.

I assume that applying it with continuity for sufficiently long time is another possible way for enhancing neural connections through neuroplasticity. Based on personal experiments performed with small groups of geoscientists (Dell'Aversana et al., 2016), practicing audio–video interpretation of seismic signals, for a period of few days, is sufficient for improving significantly the perception of anomalies in the data. Thus, *I do the hypothesis* that this type of interpretation approach can have some effect in the neural connections. At moment, there is not any neuroimaging evidence proving this hypothesis; however, it is reasonable for several reasons.

First, interactive audiovideo display of geophysical data allows activating the brain areas of hearing and vision at the same time. That approach of data analysis requires multisensory integration by the nervous system. A coherent representation of a geophysical object (such as a specific type of seismic signal, an amplitude anomaly, and so forth) combining sensory modalities is possible only if our brain creates the necessary connections between the different cortex areas. We have seen that neural structures implicated in multisensory integration are the superior colliculus and various cortical structures such as the superior temporal gyrus and visual and auditory association areas. These areas have extensive connections to each other. Furthermore, they are connected to higher association areas that are believed to integrate sensory input from various modalities. I assume that at least one type of neural connectivity (anatomical, functional, or effective) can be enhanced by multimodal–multisensory analysis of geophysical data. This intriguing work hypothesis could be confirmed or not by dedicated experiments supported by brain imaging techniques.

Another important aspect of neuroplasticity supported by multisensory analysis is selective attention. In fact, the activity of interactive audiovideo display of seismic data pushes the geophysicist to focus his/her attention every time he/she hears some anomalous pattern of sounds. These can be, for instance, trends of decreasing pitch, ensembles of musical notes characterized by high pitch and high "velocity" (sound intensity), particular notes or chords emerging from the background, sudden melodic variations, and so forth. I have noted that sometimes these particular sound patterns reveal seismic features that escaped from a first visual analysis, enhancing the presence of small faults, low-scale sedimentary features, or other weak anomalies (Dell'Aversana, 2014).

Finally, Doidge remarks that when we do an activity that requires the simultaneous activation of many specific neurons, there is a significant release of BDNF (Doidge, 2007). Thus, when we use visual and audio cortices in geophysical data analysis simultaneously, we create the optimal neural conditions for supporting our neural growth and the efficiency of synapses.

10.5.3 Field Geology and Neuroplasticity

One of the main peculiarities of the geological activity is that geologists and geophysicists often walk across long distances for exploring and interpreting a complex landscape. They observe, map, correlate heterogeneous information; imagine the subsoil; visualize in mind events happened million years ago; and try to link the pieces of the puzzle all together into a coherent geological model. This field activity allows developing unique cognitive abilities; indeed, geoscientists often move in the same scene that they are trying to interpret. In other words, they are *inside* the natural system under study. This represents an unusual point of view with respect to many other sciences, where the object of observation appears clearly disjoined from the observer.

It is possible that such an "anomalous" subject–object cognitive dynamic relationship can support neuroplasticity effects in geoscientists. In fact, *imaging-while-moving* or *interpreting-while-walking* are activities that require the combination of many cognitive processes involving different brain areas at the same time. These are, for instance, the visual cortex, sensorimotor cortex, prefrontal lobes, and so forth.

Furthermore, the DA pathways are activated during field exploration, with the effect of consolidating neural plasticity changes. Indeed, recent studies confirm that dopamine levels rise significantly by movement and physical exercise. Furthermore, aerobic exercise increases the production of neurotrophic factors including the BDNF, insulin-like growth factor 1, and vascular endothelial growth factor (Gomez-Pinilla and Hillman, 2013).

In summary, the geoexploratory activity across a complex landscape, performing mapping and imagery while moving, pattern recognition, and integration of sparse observations, all create an optimal combination of factors promoting and enhancing neuroplasticity.

10.6 SUMMARY AND FINAL REMARKS

In this chapter, I discussed the fascinating concept of neuroplasticity, showing its possible relationships with exploration geosciences. I have to admit that if there is any positive effect of the exploratory activity on our brain functionalities, it is actually unplanned and unintentional. Geologists and geophysicists have a pronounced practical sense: when they explore the environment and the Earth interior, they search for some specific target; it is rare that they think about the functioning of their neurons.

However, in this chapter, I suggested to expand our traditional view of geosciences introducing an intriguing hypothesis: the study and the practice of geological/geophysical exploration can trigger, promote, and improve neuroplasticity itself. In other words, I suggested that the activity of exploration geosciences, eventually supported by BBTs, could enhance the functionalities of our brain through neuroplastic changes. Indeed, integrating multiphysical data/models could enhance neural connections and create new synapses; interactive procedures based on hybrid visual—audio recognition could bridge different areas of our brain, activating neural growth factors; the ability to detect significant patterns in a complex background can stimulate selective attention, creating new neural links. Finally, activating the seeking system and the DA pathways through the exploratory activity in field can promote and enhance all the mechanisms of neurogenesis and synaptogenesis.

References

Arvidsson, A., Collin, T., Kirik, D., Kokaia, Z., Lindvall, O., 2002. Neuronal replacement from endogenous precursors in the adult brain after stroke. Nat. Med. 8 (9), 963—970. http://dx.doi.org/10.1038/nm747.

Bach-y-Rita, P., 1967. Sensory plasticity. Acta Neurol. Scand.

Barde, Y.A., Edgar, D., Thoenen, H., 1982. Purification of a new neurotrophic factor from mammalian brain. EMBO J. 1, 549—553.

Broca, P., 1861. Remarques sur le siege de la faculté du langage articulé, suivies d'une bservation d'aphémie. Bull. Soc. Anat. 6, 330—357.

Copi, A., Jüngling, K., Gottmann, K., December 2005. Activity- and BDNF-induced plasticity of miniature synaptic currents in ES cell-derived neurons integrated in a neocortical network. J. Neurophysiol. 94 (6), 4538–4543.

Dell'Aversana, P., Gabbriellini, G., Amendola, A., June 2016. Sonification of geophysical data through time-frequency transforms. Geophys. Prospect.

Dell'Aversana, P., 2014. A bridge between geophysics and digital music. Applications to hydrocarbon exploration. First Break 32 (5), 51–56.

Doidge, N., 2007. The Brain that Changes Itself: Stories of Personal Triumph from the Frontiers of Brain Science. Viking, New York, ISBN 978-0-670-03830-5.

Elbert, T., Pantev, C., Wienbruch, C., Rockstroh, B., Taub, E., October 13, 1995. Increased cortical representation of the fingers of the left hand in string players. Science 270 (5234), 305–307. American Association for the Advancement of Science. http://www.jstor.org/stable/2888544.

Gerstein, G.L., Perkel, D.H., 1969. Simultaneously recorded trains of action potentials: analysis and functional interpretation. Sci. New Ser. 164.

Gomez-Pinilla, F., Hillman, C., 2013. The influence of exercise on cognitive abilities. Compr. Physiol. 3 (1), 403–428. http://dx.doi.org/10.1002/cphy.c110063.

Hebb, D.O., 1949. The organization of behavior: a neuropsychological theory. John Wiley & Sons, ISBN 978-0-471-36727-7.

Hyde, K.L., Lerch, J., Norton, A., Forgeard, M., Winner, E., Evans, A.C., Schlaugc, G., 2009. The effects of musical training on structural brain development. A longitudinal study, the neurosciences and music III: disorders and plasticity. Ann. N.Y. Acad. Sci. 1169, 182–186. http://dx.doi.org/10.1111/j.1749-6632.2009.04852.x.

James, W., 1980. The Principles of Psychology. Harvard University Press, ISBN 0-674-70625-0.

Jai, T.M., 2003. Dopamine: a potential substrate for synaptic plasticity and memory mechanisms. Prog. Neurobiol. 69, 375–390.

Jenkins, W.M., Merzenich, M.M., Ochs, M.T., Allard, T., Guic-Robles, E., 1990. Functional reorganization of primary somatosensory cortex in adult owl monkeys after behaviorally controlled tactile stimulation. J. Neurophysiol. 63, 82–104.

Kleim, J.A., Jones, T.A., February 2008. Principles of experience-dependent neural plasticity: implications for rehabilitation after brain damage. J. Speech Lang. Hear. Res. 51 (1), S225–S239. http://dx.doi.org/10.1044/1092-4388(2008/018).

Kuo, M.F., Grosch, J., Fregni, F., Paulus, W., Nitsche, M.A., December 26, 2007. Focusing effect of acetylcholine on neuroplasticity in the human motor cortex. J. Neurosci. 27 (52), 14442–14447.

Lashley, K., 1923. The behaviouristic interpretation of consciousness. Psychol. Rev.

Levi-Montalcini, R., Hamburger, V., March 1951. Selective growth stimulating effects of mouse sarcoma on the sensory and sympathetic nervous system of the chick embryo. J. Exp. Zool. 116 (2), 321–361.

Merzenich, M.M., Sur, M., Nelson, R.J., Kaus, J.H., 1981. The organization of the Si cortex. Multiple representations of the body in primate, cortical sensory organization. In: Woolsey, C.N. (Ed.), 1: Multiple Somatic Areas (Listed in the bibliography for Abstract 11:965): 303.

Merzenich, M.M., Kaas, J.H., Wall, J.T., Nelson, R.J., Sur, M., Felleman, D.J., 1983a. Topographic reorganization of somatosensory cortical areas 3b and 1 in adult monkeys following restricted deafferentation. Neuroscience 8, 33–55. http://dx.doi.org/10.1016/0306-4522(83)90024-6 (Listed in the bibliography for Abstract 11:965): 303. PMID: 6835522.

Merzenich, M.M., Kaas, J.H., Wall, J.T., Nelson, R.J., Sur, M., Felleman, D.J., 1983b. Progression of change following median nerve section in the cortical representation of the hand in areas 3b and 1 in adult owl and squirrel monkeys. Neuroscience 10, 639–665. http://dx.doi.org/10.1016/0306-4522(83)90208-7 (Listed in the bibliography for Abstract 11:965). PMID: 6646426.

Merzenich, M.M., Jenkins, W.M., Middlebrooks, J.C., 1984. Observations and Hypotheses on Special Organizational Features of the Central Auditory Nervous System. In: Edleman, G., Cowan, M., Gall, E. (Eds.), Dynamic Aspects of Neocortical Function. John Wiley and Sons, New York (Listed in the bibliography for Abstract 11:965): 303.

Panksepp, J., Biven, L., 2012. The archaeology of mind: neuroevolutionary origins of human emotions. Nort. Ser. Interpers. Neurobiol.

Rasmusson, D.D., November 2000. The role of acetylcholine in cortical synaptic plasticity. Behav. Brain. Res. 115 (2), 205–218.

Recanzone, G.H., Merzenich, M.M., Jenkins, W.M., Grajski, K.A., Dinse, H.R., 1992. Topographic reorganization of the hand representation in cortical area 3b owl monkeys trained in a frequency-discrimination task. J. Neurophysiol. 67, 1031–1056.

Sarter, M., Bruno, J.P., Givens, B., November 2003. Attentional functions of cortical cholinergic inputs: what does it mean for learning and memory? Neurobiol. Learn. Mem. 80 (3), 245–256.

Shatz, C.J., 1992. The developing brain. Sci. Am. ISSN: 0036-8733 60–67.

Siegel, D.J., 2012. Pocket Guide to Interpersonal Neurobiology: An Integrative Handbook of the Mind, Mind Your Brain, Inc.

Skudlarskia, P., Jagannathana, K., Calhouna, V.D., Hampsone, M., Skudlarskaf, B.A., Pearlsona, G., November 15, 2008. Measuring brain connectivity: diffusion tensor imaging validates resting state temporal correlations. NeuroImage 43 (3), 554–561.

Sporns, O., 2007. Brain connectivity. Scholarpedia 2 (10), 4695.

Veenstra-VanderWeele, J., Anderson, G.M., Cook, E.H.M., 2000. Pharmacogenetics and the serotonin system: initial studies and future directions. Eur. J. Pharmacol. 410 (2–3), 165–181.

Epilogue

Exploration can be defined as "the act of searching information or resources." This definition is intuitive and implies both physical and cognitive activities; thus, I started from it, at the beginning of this book, for introducing the discussion about the neurobiological background of exploration geosciences.

At the end of this long dissertation, I have all the elements for expanding that definition into a more comprehensive and deeper idea of exploration. This expanded view includes other concepts, such as cognition, emotions, significance, semantic process, organized and disorganized complexity, creativity, neuroplasticity, and brain empowerment. Apparently, all these concepts are not directly linked with the practical activities of geologists and geophysicists. Nevertheless, they are.

In 1948, the American mathematician Warren Weaver remarked the difference between *disorganized and organized complexity*. The first case is referred to systems characterized by many independent variables, so that they can be described through sophisticated statistical analysis. Typical examples are the thermodynamic systems, which can be treated using probability theory and statistical mechanics. Instead, "organized complexity" deals with phenomena and systems formed by a "sizable number of parts, which are interrelated into an organic whole" (Weaver, 1948). They cannot be analyzed just by statistical approaches, because we cannot neglect to take into account the correlations and the interactions between the constituent parts. An example of system with organized complexity is our brain, where billions of neurons are connected through trillions of synapses and work for the same adaptive biological objectives.

This impressive neurobiological organization allows us transforming multisensory perception of heterogeneous data into coherent concepts, significances, models, and theories. I used the expression "semantic process" for indicating this path that goes from disorganized informative complexity to organized cognitive complexity. Seeking for food and refuge, exploring the subsoil for discovering a hydrocarbon reservoir, imaging the body interior for doing a medical diagnosis, all represent different types of the semantic process.

In this book, I have remarked many times that this process involves both cognitive and emotional aspects. In fact, significances emerge progressively from the continuous interaction between subcortical nuclei and neocortex. That interaction, together with the continuous feedback from the environment, allows "the mind" performing complex inductive inferences. These represent the main peculiarity of biological intelligence, in human beings as well as in many other animal species. Reproducing the inferential process with a machine represents actually one of the main scientific challenges.

The inferential and the semantic processes are particularly complex in the human brain. One reason is that they commonly involve complicate relationships within large communities. When also scientific approaches and advanced technologies are used, the whole process is still more complex. Sometimes, that complexity overload is above our possibility to manage. In that case, we can fail to combine huge information streams (big data) into coherent semantic structures. The dramatic consequence can be the informative/semantic chaos. We can lose the capability to take decisions, to understand the events happening around us, and to move forward.

However, if we have enough resources in terms of energy, technology, people, space, time, relationships, motivations, emotions, and expectations, then we can face the chaos. If we are lucky and clever enough, we can restore a condition of (at least partial) organized complexity. This result can be obtained, for instance, through proper selection and integration of the redundant information streams. The positive consequences will be some type of creative result, such as a new idea, a new theory, a new model, and a new discovery. In other words, we can use the complexity and the chaos for being creative.

This creative process often happens in exploration geosciences. For instance, the new challenges of hydrocarbon exploration happening in more and more difficult geological contexts offer the chance to geoscientists and managers to be creative. Thus, exploring the Earth's interior can represent a unique opportunity to have new ideas, to invent new technologies, to discover new resources. Finally, I guess, it is also a great opportunity for all of us to empower our brain.

In conclusion, we can see human exploration as a challenging adventure of our creativity through the complexity. This is not only a material travel through the physical space. It is mainly a cognitive and emotional path through the mind. Every type of new discovery is neither just a technical nor a commercial success. It is also a cognitive result emerging in the complexity of our brain along the semantic path from disorganized to organized complexity.

Reference

Weaver, W., 1948. Science and complexity. Am. Sci. 36 (4), 536–544. PMID: 18882675.

Appendix 1: Overview About Inversion

Let us assume that we know the parameters of a model (physical, mathematical, or any other type of model). For instance, it can be an Earth model in terms of distribution of compressional seismic velocity. If we have a mathematical relationship linking the model parameters with some type of observable quantity, we can perform a simulation and predict the theoretical values of those observables. In the case of the seismic velocity model, for instance, we can simulate a seismic experiment and calculate the travel times of seismic waves observed at an array of geophones. We call this prediction a "simulated response." The problem of simulating the response for an assigned model is called a *forward problem*. The opposite question is called an *inverse problem*. In the previous example, it corresponds to the inversion of the experimental seismic response (travel times) to retrieve a model of the geophysical parameters (seismic velocity field). In a general sense, solving the inverse problem involves using the actual result of some measurements to infer the values of the parameters that characterize the system (Tarantola, 2005).

In this appendix, I summarize the fundamentals of geophysical inversion: estimating models of physical property distributions based on geophysical survey data.

The measurements in a given geophysical domain are commonly represented by a vector named *data vector*:

$$\vec{d} = [d_1, ..., d_i, ..., d_{N-1}, d_N]^{\text{T}}. \tag{A1.1}$$

The model parameters are commonly represented by the model vector:

$$\vec{m} = [m_1, ..., m_i, ..., m_{M-1}, m_M]^{\text{T}}. \tag{A1.2}$$

The superscript "T" in Eqs. (A1.1) and (A1.2) means "transposition."

The data and model vectors are linked by physical relationships represented by an operator g, generally nonlinear, commonly called the *forward operator*:

$$g(\vec{m}) = \vec{d}. \tag{A1.3a}$$

If the relation for solving the forward problem is linear we can write

$$G\vec{m} = \vec{d}. \tag{A1.3b}$$

An alternative notation frequently used is

$$\mathbf{Gm} = \mathbf{d}, \tag{A1.4}$$

where the bold notation indicates that the forward operator can be expressed by its representative matrix and that data and model parameters can be expressed as vectors (Menke, 1989).

The linear inverse problem can be formulated as

$$\mathbf{m} = \mathbf{G}^{-1}\mathbf{d}. \tag{A1.5}$$

For instance, we could have a set of $M > 2$ experimental measurements of temperature T_i versus depth Z_i:

$$T_i = m_1 + m_2 Z_i. \tag{A1.6}$$

In this case, the inverse problem is said to be *overdetermined*. In general, it has no exact solution. However, we can search for a least squares solution representing the best estimate of the model vector $\mathbf{m} = \begin{bmatrix} m_1 & m_2 \end{bmatrix}^{\mathrm{T}}$.

That solution is (Menke, 1989)

$$\mathbf{m}^{\mathrm{est}} = \begin{bmatrix} \mathbf{G}^{\mathrm{T}}\mathbf{G} \end{bmatrix}^{-1} \mathbf{G}^{\mathrm{T}}\mathbf{d}. \tag{A1.7}$$

In many experimental sciences, the relation solving the forward problem is not linear. The weakest case of nonlinearity arises when the function $g(m)$ can be linearized around an *a priori* model, $\mathbf{m}_{\mathrm{prior}}$. In this case, neglecting terms higher than the first order, we can write

$$g(\mathbf{m}) \cong g(\mathbf{m}_{\mathrm{prior}}) + \mathbf{J}(\mathbf{m} - \mathbf{m}_{\mathrm{prior}}), \tag{A1.8}$$

where $J_\alpha^i = \left(\dfrac{\partial g^i}{\partial m^\alpha} \right)_{\mathbf{m}_{\mathrm{prior}}}.$

Moving to a stochastic approach,[1] in the least squares formulation[2] of the inverse problem, the (twice) objective function to minimize is given by (Tarantola, 2005)

[1] In probability theory, a stochastic system (as well as a stochastic process) is characterized by nondeterministic states (for a detailed description and applications in inversion of geophysical data, see Tarantola, 2005).

[2] The method of least squares is a standard approach to the approximate solution of overdetermined systems. Where the observations have uncertainties with normal (Gaussian) distribution, the least squares method corresponds to the maximum likelihood criterion.

$$2S(\mathbf{m}) = \|g(\mathbf{m}) - \mathbf{d}_{\text{obs}}\|_D^2 + \|\mathbf{m} - \mathbf{m}_{\text{prior}}\|_M^2$$
$$= (g(\mathbf{m}) - \mathbf{d}_{\text{obs}})^{\mathrm{T}} \mathbf{C}_D^{-1} (g(\mathbf{m}) - \mathbf{d}_{\text{obs}})$$
$$+ (\mathbf{m} - \mathbf{m}_{\text{prior}})^{\mathrm{T}} \mathbf{C}_M^{-1} (\mathbf{m} - \mathbf{m}_{\text{prior}}). \tag{A1.9}$$

$\mathbf{C}_D$ and $\mathbf{C}_M$ are, respectively, the covariance data matrix and covariance model matrix. It is possible to verify that the *a posteriori* probability density is approximately Gaussian, with a center given by

$$\mathbf{m}^{\text{est}} \cong \mathbf{m}_{\text{prior}} + \left(\mathbf{J}^{\mathrm{T}}\mathbf{C}_D^{-1}\mathbf{J} + \mathbf{C}_M^{-1}\right)^{-1}\mathbf{J}^{\mathrm{T}}\mathbf{C}_D^{-1}(\mathbf{d}_{\text{obs}} - g(\mathbf{m}_{\text{prior}})). \tag{A1.10a}$$

The corresponding *a posteriori* covariance operator, which describes the uncertainty of the estimated model, is given by (Tarantola, 2005)

$$\mathbf{C}_M^{\text{post}} = \left(\mathbf{J}^{\mathrm{T}}\mathbf{C}_D^{-1}\mathbf{J} + \mathbf{C}_M^{-1}\right)^{-1} = \mathbf{C}_M - \mathbf{C}_M\mathbf{G}^{\mathrm{T}}\left(\mathbf{J}\mathbf{C}_M\mathbf{J}^{\mathrm{T}} + \mathbf{C}_D\right)^{-1}\mathbf{J}\mathbf{C}_M. \tag{A1.10b}$$

We can use an iterative approach to obtain the maximum likelihood estimate of the model vector. This means that we start from an initial model (starting guess) and then we update it iteratively.

For instance, using quasi-Newtonian optimization approach, the iterative formula is

$$\mathbf{m}_{n+1} = \mathbf{m}_n - \mu_n \left(\mathbf{J}_n^{\mathrm{T}}\mathbf{C}_D^{-1}\mathbf{J}_n + \mathbf{C}_M^{-1}\right)^{-1}\left[\mathbf{J}_n^{\mathrm{T}}\mathbf{C}_D^{-1}(\mathbf{d}_n - \mathbf{d}_{\text{obs}})\right.$$
$$\left. + \mathbf{C}_M^{-1}(\mathbf{m}_n - \mathbf{m}_{\text{prior}})\right]. \tag{A1.11}$$

The model update is driven by the misfit $(\mathbf{d}_n - \mathbf{d}_{\text{obs}})$, which is back propagated with every iteration. Here, $\mathbf{d}_n$ represents the predicted response at iteration n. The parameter $\mu_n \leq 1$ is "an ad hoc parameter defining the size of the jump to be performed at each iteration" (Tarantola, 2005).

Back propagation means that all the model parameters influencing the response are "adjusted" as a function of the misfit between observations and predictions. For instance, in the case of seismic travel-time inversion, the model parameters are the seismic velocities and the misfit is the difference between the observed and predicted travel times. Different algorithms can be used to optimize the model updating, such as Newtonian, quasi-Newtonian and conjugate gradient methods.

References

Menke, W., 1989. Geophysical Data Analysis: Discrete Inverse Theory. Academic Press.
Tarantola, A., 2005. Inverse problem theory and methods for model parameter estimation. Soc. Ind. Appl. Math.

Appendix 2: Overview About Simultaneous Joint Inversion

In simultaneous joint inversion (SJI), complementary methods are combined in a single inversion scheme honoring all data simultaneously (Dell'Aversana, 2014). This objective is obtained by minimizing a "joint cost functional" including two or more misfit terms, plus one or more regularization terms. Furthermore, some type of constraint and relationship between the different model parameters are included into the functional (De Stefano et al., 2011).

In this appendix, I provide a brief summary of SJI addressed to geophysical problems. However, the same general formulation can be applied for jointly inverting many other categories of data, including measurements performed on human body. Thus, SJI is an integrated approach suitable for multiparametric medical imaging too.

In geophysics, data belonging to different and *complementary* domains can be inverted simultaneously with the scope of obtaining models of different, but correlated, geophysical (or rock physics) parameters. For instance, seismic and electromagnetic data are jointly inverted to obtain a model of porosity and fluid saturation in a hydrocarbon reservoir (Dell'Aversana et al., 2011).

In SJI, different types of theoretical responses must be compared with the different types of observations in each specific geophysical domain. Thus, a joint misfit function is created, including the differences between observed and predicted quantities in the various domains. Furthermore, we need to include the relationships between the different model parameters to constrain the inversion process by physical or statistical links.

For instance, let us suppose that we desire to invert seismic and electromagnetic data simultaneously for retrieving a model of porosity and saturation in a hydrocarbon reservoir. We must use a rock physical relationship that links both seismic and electromagnetic measurements with porosity and saturation. That relationship must be taken into account in the SJI scheme.

This approach can be formalized through the minimization of a joint objective functional including the joint misfit function, the relationships between the parameters and one or more regularization terms.

Following the stochastic approach of Tarantola (2005), the misfit functional for a given inverse problem assigned in a certain geophysical domain is defined as

$$\Phi_{\mathrm{mis}}(\mathbf{m}) = [g(\mathbf{m}) - \mathbf{d}_{\mathrm{obs}}]^{\mathrm{T}} \mathbf{C}_d^{-1} [g(\mathbf{m}) - \mathbf{d}_{\mathrm{obs}}], \qquad (A2.1)$$

where $\mathbf{m}$ and $\mathbf{d}_{\mathrm{obs}}$ are the *multimodel* and the *multidata* vectors, respectively. Two or more types of parameters and measurements form the multimodel and the multidata vectors, respectively. For instance, $\mathbf{m} = [\mathbf{m}_1, \mathbf{m}_2, \mathbf{m}_3]$, where $\mathbf{m}_1$ is the vector of seismic velocity, $\mathbf{m}_2$ is the vector of electric resistivity, and $\mathbf{m}_3$ is the vector of density distribution.

Analogously, $\mathbf{d} = [\mathbf{d}_1, \mathbf{d}_2, \mathbf{d}_3]$, where $\mathbf{d}_1$ is (for instance) the vector of observed seismic travel times, $\mathbf{d}_2$ is the vector of electric potentials, and $\mathbf{d}_3$ is the vector of Bouguer anomalies.

$g(\mathbf{m})$ is the simulated response calculated using the forward operator g (nonlinear in the most general case), and $\mathbf{C}_d^{-1}$ is the inverse of the data covariance matrix.

We need to estimate a misfit function for each geophysical domain. In fact, for each domain, a specific forward operator will be applied to the correspondent model vector to calculate the theoretical response to be compared with the observations. The joint misfit function will be simply the sum or a linear combination of the individual misfit functions.

From inverse problem theory, it is well known that a regularization term is required to make the inverse problem well posed (Tikhonov and Arsenin, 1977).[1] A commonly used regularization form is based on smoothing operators (Vozoff and Jupp, 1975), like the following (Moorkamp et al., 2011):

$$\Phi_{\mathrm{reg}}(\mathbf{m}) = \sum_i \alpha_i (\mathbf{m} - \mathbf{m}_0)^{\mathrm{T}} W_i^{\mathrm{T}} \mathbf{C}_M^{-1} W_i (\mathbf{m} - \mathbf{m}_0)$$

$$+ \beta (\mathbf{m} - \mathbf{m}_0)^{\mathrm{T}} \mathbf{C}_M^{-1} (\mathbf{m} - \mathbf{m}_0), \qquad (A2.2)$$

where $i = \{x, y, z\}$ represents the three spatial directions and α_i is a weighting factor; $\mathbf{m}_0$ is a reference model; $\mathbf{C}_M^{-1}$ is the inverse of the model covariance matrix; and the matrix W_i represents an approximation of the spatial (first or second) derivative of the model parameters. For instance, if this matrix represents first derivatives of model parameters, models with minimum variations of the parameter between adjacent cells will result from the inversion. The second term in (A2.2), including the weighting factor β, allows us regulating how much we desire to take the actual model close to our reference model. β can be fixed, but more

---

[1] An ill-posed inverse problem is characterized by the fact that small error bars in the observations are reflected as large oscillations in the models. Different types of regularization can be used for stabilizing the inversion results.

frequently it can change during the inversion, to make the inverse problem more or less model driven.

In many cases, we can reasonably assume that different model parameters change consistently in three-dimensional space. This assumption can be formalized in terms of structural similarity constraints and can be included as an additional term in the objective function. Gallardo and Meju (2003, 2004, 2007) describe this approach by introducing cross-gradient terms in the form:

$$\Phi_X(\mathbf{m}) = (\nabla\mathbf{m}_1 \times \nabla\mathbf{m}_2)^T \mathbf{C}_M^{-1}(\nabla\mathbf{m}_1 \times \nabla\mathbf{m}_2). \tag{A2.3}$$

Here, $\mathbf{m}_1$ and $\mathbf{m}_2$ represent model vectors describing two different, but spatially correlated, physical properties, for instance, seismic P-velocity and electric resistivity, changing according to geological variations.

De Stefano et al. (2011) discuss a "compact formulation" of the SJI problem, including a general approach for introducing empirical links between the different parameters in the objective function. For instance, including Gardner's law between seismic P-velocity v_P and density d as a constraint during the inversion is translated into the additional cost function term:

$$\Phi_{\text{Gardner}}(v_P, d) = \left| a \cdot v_P^b - d \right|^2. \tag{A2.4}$$

In a similar way, it is possible to introduce every type of "analytical" constraint based on explicit empirical or rock physics relationships between different parameters:

$$\Phi_{\text{analytic}}(\mathbf{m}_1, \mathbf{m}_2) = |f(\mathbf{m}_1) - \mathbf{m}_2|^2, \tag{A2.5}$$

where f indicates the generic functional relationship between the two generic parameters $\mathbf{m}_1, \mathbf{m}_2$.

In summary, the generic form of the joint objective function can be written as

$$\Phi_{\text{joint}} = \lambda_1 \Phi_{\text{mis}} + \lambda_2 \Phi_{\text{reg}} + \lambda_3 \Phi_X + \lambda_4 \Phi_{\text{analytic}}. \tag{A2.6}$$

The factors λ_i are used for weighting the different terms, depending on the geological context, the type and quality of the data, and other factors. How these weighting factors are effectively defined represents a fundamental question. This is often solved through empirical approaches (by trial and error or taking into account for the variable data quality and intrinsic resolution of the different data sets).

References

Dell'Aversana, P., 2014. Integrated Geophysical Models: Combining Rock Physics with Seismic, Electromagnetic and Gravity Data. EAGE Publications.

Dell'Aversana, P., Bernasconi, G., Miotti, F., Rovetta, D., 2011. Joint inversion of rock properties from sonic, resistivity and density well-log measurements. Geophys. Prospec. 59 (6), 1144–1154.

De Stefano, M., Golfré Andreasi, F., Re, S., Virgilio, M., Snyder, F.F., 2011. Multiple-domain, simultaneous joint inversion of geophysical data with application to subsalt imaging. Geophysics 76 (3), R69.

Gallardo, L.A., Meju, M.A., 2003. Characterization of heterogeneous near-surface materials by joint 2D inversion of DC resistivity and seismic data. Geophys. Res. Lett. 30 (13), 1658. http://dx.doi.org/10.1029/2003GL017370.

Gallardo, L.A., Meju, M.A., 2004. Joint two-dimensional DC resistivity and seismic travel time inversion with cross-gradients constraints. J. Geophys. Res. 109, B03311. http://dx.doi.org/10.1029/2003JB002716.

Gallardo, L.A., Meju, M.A., 2007. Joint two-dimensional cross-gradient imaging of magnetotelluric and seismic traveltime data for structural and lithological classification. Geophys. J. Int. 169, 1261–1272.

Moorkamp, M., Heincke, B., Jegen, M., Roberts, A.W., Hobbs, R.W., 2011. A framework for 3-D joint inversion of MT, gravity and seismic refraction data. Geophys. J. Int. 184, 477–493. http://dx.doi.org/10.1111/j.1365-246X.2010.04856.x.

Tarantola, A., 2005. Inverse problem theory and methods for model parameter estimation. Soc. Ind. Appl. Math.

Tikhonov, A.N., Arsenin, V.Y., 1977. Solution of Ill-Posed Problems. V.H. Winston and Sons.

Vozoff, K., Jupp, D.L.B., 1975. Joint inversion of geophysical data. Geophys. J. 42, 977–991.

Index

'*Note*: Page numbers followed by "f" indicate figures.'

Printed in the United States
By Bookmasters